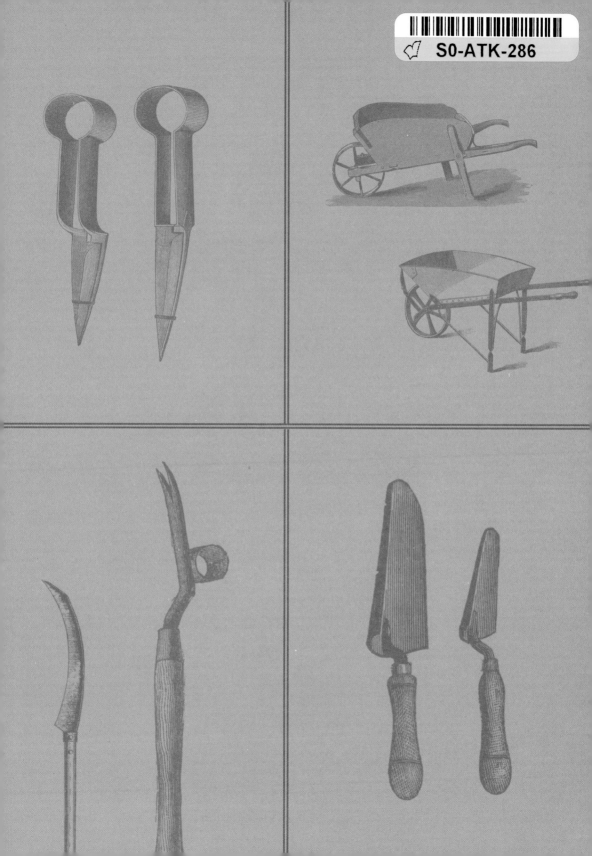

A

HISTORY

of the

GARDEN

in

FIFTY TOOLS

Bill Laws lives in Hereford, England. His other books include *Fifty Plants that Changed the Course of History*, *Fifty Railroads that Changed the Course of History*, and *The Field Guide to Fields*.

The University of Chicago Press, Chicago 60637
The University of Chicago Press, Ltd., London
© 2014 Quid Publishing
All rights reserved. Published 2014.
Printed in China

23 22 21 20 19 18 17 16 15 14 13   1 2 3 4 5

ISBN-13: 978-0-226-13976-0 (cloth)
ISBN-13: 978-0-226-13993-7 (e-book)
DOI: 10.7208/chicago/9780226139937.001.0001

Library of Congress Cataloging-in-Publication Data
Laws, Bill, author.
  A history of the garden in fifty tools / Bill Laws.
      pages ; cm
  Includes index.
  ISBN 978-0-226-13976-0 (cloth : alk. paper) — ISBN 978-0-226-13993-7 (e-book)
  1. Garden tools—History. 2. Gardening--Equipment and supplies—History. I. Title.
  SB454.8.L395 2014
  635—dc23
                                             2013042254

A

HISTORY

of the

GARDEN

in

FIFTY TOOLS

BILL LAWS

THE UNIVERSITY OF CHICAGO PRESS

Chicago and London

Contents

Introduction   6

CHAPTER ONE

The Flower Garden

Fork   10

Bulb Planter   15

Trowel   19

Hand Pruning Shears   23

Garden Basket   28

Soil-Test Kit   32

Dibber   35

Wellington Boot   39

Hat and Gloves   43

Garden Catalog   46

Garden Journal   50

CHAPTER TWO

The Vegetable Garden

Spade   56

Hoe   60

Mattock   65

String Line   68

Billhook   72

Rake   76

Mechanical Tiller   80

Composter   84

Hotbed   89

Latin   94

Raised Bed   99

Soil Sieve   104

Radio   107

CHAPTER THREE

The Lawn

Lawn mower 112

Sickle and Scythe 117

Grass and Hedge Shears 121

Daisy Grubber 125

Weed Killer 129

Fertilizer 134

Tape Measure 138

CHAPTER FOUR

The Orchard

Ladder 144

Grafting Knife 147

Pruning Saw 151

Fruit Barrel 155

Label 159

Thermometer 164

Scarecrow 168

CHAPTER FIVE

Structures & Accessories

Potting Shed 174

Glasshouse 178

Cloche 182

Terrarium 186

Plant Container 190

Terracotta Pot 195

Stoneware Urn 200

Wheelbarrow 203

Patio Brush 207

Sundial 211

Hose 214

Watering Can 218

Index 222

Picture Credits 224

Introduction

For hundreds of years, and still today, gardening has ranked as one of our most popular pastimes. Reading about gardening traditionally comes a close second and yet, while much has been written about great gardens, famous gardeners and notable plants, those instruments of gardening, the tools that root us to the soil, have received short shrift. Without our tools, we would have no gardens.

S ome of the best tales start with a garden. "The Book of Life begins with a man and a woman in a garden," declares Lord Illingworth to Mrs. Allonby in Oscar Wilde's *A Woman of No Importance* (1893). As the wise Mrs. Allonby reminds him, the Book ends "with Revelations." Every tool has a tale to tell and the anecdotes that follow bring to light a host of horticultural revelations.

There were the extraordinarily useful pruning shears invented by the 18th-century Comte Bertrand de Molleville who had so recently escaped the French Revolution's guillotine, and Dr. Nathaniel Ward's glass and brass portable glasshouse, an instrument that transformed the variety of plants available to 19th-century garden designers like Gertrude Jekyll; there was the first, two-stroke, Victa mower made by Australia's Mervyn Richard-son from an old soap box and an empty peach can, and the Hall of Fame's Charles Goodyear, who helped invent the rubber hose (leading to the spread of the lawn into southern climes) and the rubber boot (which Jekyll—"I do not like new boots"—ignored).

A good garden tool, such as the steel fork that transforms flower borders in spring or the trusty spade that tempers the allotment in winter, is more than a mere *gadget*: it is a pleasure to use. Some older garden tools are superior to their modern counterparts both in the way they were crafted and in the quality of the materials used. They still fetch a good price on the market and they will improve with age and care. Despite the author Washington Irving's assertion that the only edged tool that grows keener with constant use is a sharp tongue (*Rip Van Winkle*, 1819), a dash of oil and the touch of the whetstone will extend the working life of a well-loved tool. *A History of the*

Mary, Mary, quite contrary, how does your garden grow? Better, with the help of good garden tools.

Garden in Fifty Tools offers some useful tips not only on regular care and maintenance, but also on the best ways to use them.

Settling on which tools to include in the tool shed, and which to exclude, has been thought-provoking. Certain tools, evolving slowly down the centuries, have outlived the history of gardening itself: the mattock made its mark long before the dawn of horticulture, while the moon-shaped sickle has been keeping down the weeds since Neolithic times. The well-honed Dutch hoe, the Japanese rake or the neatly balanced French watering can could not be ignored. However, the choice of more controversial items such as Latin nomenclature, the garden journal, the sundial (described by the Victorian garden writer John Claudius Loudon as both "agreeable and useful") or even the radio has been dictated by the fact that each has a story to tell.

For convenience the 50 tools in this book are arranged according to the different parts of the garden, although few are exclusive to those areas. Most gardeners possess some well-tempered and cherished piece of gardening gear that accompanies them everywhere. My personal favorite is

Some of our older tools, such as the string line and shears, are still serviceable and continue to perform their duties as well today as they ever did.

a 30-year-old Opinel pocket knife; my nonagenarian mother's somewhat eccentric choice is a plastic bucket with drainage holes drilled in the base. And woe betide the borrower who fails to return that much-loved tool, or worse still brings it back broken or damaged: "Few lend (but fools) their working tools," noted the 16th-century poet and farmer Thomas Tusser in his *Five Hundred Points of Good Husbandry.*

The spade, the rake and the grafting knife, illustrated here in Abel Grimmer's 1607 Der Herbst, *also have a place in the contemporary garden shed. And each has a tale to tell.*

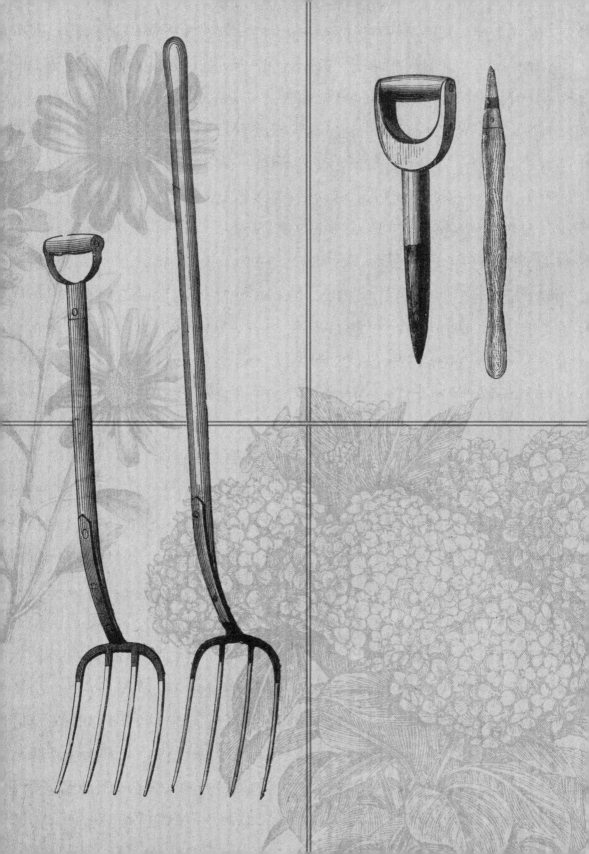

The Flower Garden

Digging around in the flower borders requires a light touch, a detailed record of what is planted and where (bring out the garden journal), and a variety of special tools. The garden fork, however, ranks as one of the most useful.

Fork

As the most widely used tool in the
garden, the fork has appeared in
many forms: flat-tined garden or
spading forks, two-handled broadforks,
four-tine manure forks and pitch, or
hay, forks among them. It is a strong
contender to be the most useful tool
in the garden.

Definition

A multi-use tool for turning soil, aerating
lawns, dividing perennials and moving
compost and manure.

Origin

Although the name originates from the
Latin *furca*, the implement was a relative
latecomer to the garden shed.

T he fork comes in many shapes and sizes: there are two-tined forks for lifting parsnips without gashing them; fork-hoe combination tools for working quickly around flower borders between the buried bulbs; and sharp-ended, scoop-shaped forks for gathering up manure or compost. Forks designed to be used on heavy clay soils tend to have straight tines to provide better leverage, whereas slightly curved tines suit lighter soils. While North Americans are happier with their D-shaped handle grips, this design frustrates gardeners with large hands: like the traditionalists of northern England, they prefer a T-shaped grip.

Many gardeners left their fork on the table and managed without in the garden: back in 1657 an inventory of items owned by the governor of the New Haven colony in Connecticut, one Theophilus Eaton, listed garden shears, sickles, hooks, hoes, "sithes," stone axes, brick axes and a trowel, but neither fork nor spade. One reason may have been the indifferent quality of metal available at the time. A fork was a tricky tool to make, while a long-handled shovel was much simpler and more reliable.

THIEF-PUNISHER

One theory on the origins of the word *fork* suggests a derivation from the Latin *fur*, a thief, as "originally the instrument for punishing thieves" (*Gresham English Dictionary*, 1931). Be that as it may, the Romans were industrious in throwing out their old, worn-out wood and bone tools, and replacing them with iron substitutes. And, once they had occupied the Celtic kingdom they called Noricum (which now forms part of Austria and Slovenia), they had access to the notoriously hard Noric steel. Understandably, the Romans devoted most of their steelworking skills to the manufacture of the *gladius*, the sword that helped them subjugate Europe. Yet out of swords came plowshares and garden tools. The Romans who first moored their boats on the harborside at Londinium in England around A.D. 43 not only brought their steel swords with them, but their ideas on civic buildings, roads, villas, villa gardens and even the flowers and vegetables that were to grow in them. By this time the wild cabbage (*Brassica oleracea*) had been tamed, and the Romans introduced its domesticated cousin into northern Europe along with their *pastinaca* (parsnip), radish, lettuce and

Garden forks, from the D-handled, flat-tined, three-pronged fork on the left to the four-pronged manure fork on the right, were essential aids to horticulture.

turnip. Once they had defeated the woad-painted British warriors, they also introduced their farm and garden tools: mattocks and metal-shod wooden spades, adzes and sickles, and the three-pronged fork. A model of such a fork was found in a cache in a tomb on the Rhine.

In medieval times, from the 5th to the 15th centuries, there is scant evidence of garden tools used. While the 13th-century illustrated French Gothic manuscript, *Très Riches Heures du Duc de Berry*, shows a maidservant piling hay with her pitchfork, a 14th-century English manuscript, the *Luttrell Psalter*, illustrates wooden weed-pulling hooks and tongs, but no forks. Yet a tool with such a variety of uses must have been widely employed in the herbers and among the raised beds of vegetable gardens. By the 1650s, when the diarist John Evelyn began writing his magnum opus, the *Elysium Britannicum* (still unfinished at his death in 1706), the grand Renaissance gardens of Italy, France and the Low Countries had begun to influence garden makers in other parts of Europe. Evelyn illustrates many of the garden tools of the time: they include a plant house that resembles a four-poster bed, great contraptions for drawing a grass roller up a slope, a massive mobile tree planter and—alongside some metal-shod wooden spades and a pickaxe— a fork, its three-pronged head apparently made of iron and fixed to a wooden handle. The garden fork as we know it had arrived.

John Evelyn the English gardener and diarist, sketched a range of tools for his 17th century Elysium Britannicum. The garden instruments included a two-spout water pot, several cold frames and a four-poster bed used to raise plants.

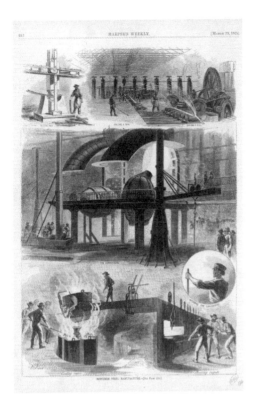

Metal manufacturing and new steelworks such as Henry Bessemer's would transform the quality of garden tools during the Industrial Age.

The fork did not change much until the advent of the alchemists of the Industrial Age—inventive men such as Henry Bessemer, who came up with a method of mass-producing steel that represented the biggest breakthrough in tool technology since the fall of the Roman empire. But, design-wise, the usually inventive Victorians added little to the stock of hand tools; the villa gardener of Roman times could have reliably recognized most of what hung in the 19th-century tool shed. An innovative steel fork made by Mr. Alexander Parkes of Birmingham was shown at the Great Exhibition at London's Crystal Palace

in 1851. The gardener James Shirley Hibberd was a convert: "Let me commend the steel digging-forks that are now getting into such general use. For all ordinary digging they are better than spades, because they break the soil well, pass through it easily, and enable a man to perform one-third more work in a day than with a spade," he would declare in his book *Profitable Gardening* (1863).

THE FORKLESS GARDEN

But a century or so later there were those who neither welcomed nor wanted a fork in the garden. In 1973 the oil producers of the Middle East, in a dispute with the West, placed an embargo on petroleum supplies. The resulting oil crisis caused a sober reappraisal of attitudes to fuel, food and sustainability. This green thinking was given added impetus by evidence, published in the 1990s but still disputed by some, of potentially disastrous climate change. One of the practical and philosophical reactions was the idea of "permaculture." This term, combining the attributes of permanence and cultivation, offered a sustainable and holistic approach to a range of gardening techniques.

As one of its leading exponents, the Tasmanian biologist and environmentalist Bill Mollison, put it, this was a philosophy of "working with, rather than against nature . . . of looking at plants and animals in all their functions, rather than treating any area as a single product system."

One of its practical manifestations was the no-dig vegetable bed. The idea of leaving the land to take care of itself—of performing no digging and applying no compost, chemical fertilizer or herbicide—was promoted by the Japanese farmer Masanobu Fukuoka, who had spent his early career as a soil scientist. Fukuoka, who died in 2008, used neither hoe nor fork on his fruitful little family farm on the island of Shikoku. Other forms of no-dig gardening involved clearing the growing area of weeds then covering the ground with a generous layer of compost and broadcast-sowing the seeds. From that point on the garden fork was needed only to haul out the humus from the compost pile and carry it to the vegetable bed. It was an appropriate time to reintroduce an old design: the right-angled fork, sometimes called a *graip* (possibly from the Danish *greb*) and resembling a fork-sized rake. Also known as a muck, or manure, fork, it was used to pull or "grab" material from the pile. These heaps were often raided by nearby gardeners before they were carted away: containing small stones and well-rotted leaf mould, they were a free source of potting compost (see p. 84).

TOOLS IN ACTION
Fork Care

A good garden fork can last a lifetime, albeit with a replacement handle from time to time; a cheap garden fork, which will last a season or two, is not worth the trouble. Street markets and yard sales are an invaluable source of sound old garden forks. Check for bent tines and woodworm (tiny but regular bug-sized holes drilled in the woodwork), but do not be put off by either. A bent tine can often be straightened by setting it in a metal scaffolding tube and gently levering down on it. Handles with some woodworm should be treated with a proprietary woodworm killer, or if the damage is substantial, replaced.

The no-dig garden lobby brought another old design back into fashion: the Basque *laya*, a multi-pronged fork set along the base of a two-handled frame. It is used to fork over the soil and, according to its advocates, it is an ergonomically perfect tool.

The fork ... which is acted on like the spade, by means of a shoulder or hilt, for thrusting it into the matters to be forked; and as a lever or handle for separating and lifting them.

John Claudius Loudon, *An Encyclopaedia of Gardening* (1822)

Bulb Planter

Requiring neither a power source nor an instruction manual, the bulb planter is the epitome of sustainable simplicity: screw it into the ground to remove a plug of soil, drop in the bulb and replace the soil. Bulb planters have had an especially important role to play with the rise of the wild garden.

Definition

A long- or short-handled implement with a hollow core, used to make planting holes for flowering bulbs.

Origin

Fairly recent in origin, bulb planters and turf pluggers gained popularity in the 20th century.

T his is a no-fuss tool, so much so that some gardeners regard it as unnecessary to the task at hand: creating a hole in the turf or border in order to plant a bulb. However, hours spent hand-planting a sack of narcissi in a sweep of old turf will persuade anyone to go out and buy one. The early 19th-century garden guru John Claudius Loudon believed in their usefulness. Trowel-planting,

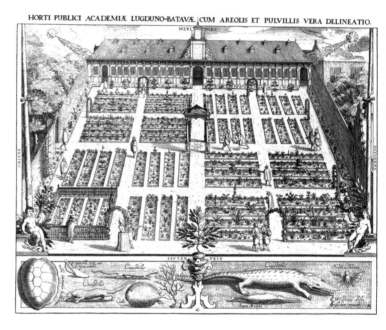

HORTI PUBLICI ACADEMIÆ LUGDUNO-BATAVÆ, CUM AREOLIS ET PULVILLIS VERA DELINEATIO.

Charles de l'Écluse, director of the botanical garden at the University of Leiden (above in 1610) introduced the tulip bulb to Europe. The bulb planter was not far behind.

he wrote in his *Encyclopaedia of Gardening* (1822), is performed with a garden trowel "made hollow like a scoop." The concept of a bulb planter, however, which he conjectured had originated in France, had been improved upon by a Mr. Saul who produced an alarming, speculum-like instrument with a hinged handle.

It is more likely that the bulb planter started out in the Low Countries, where nurserymen had profited from the business of bulbs since the 1630s. This was a time when tulip bulbs were traded like expensive paintings, thanks to the collection of the Flemish botanist Charles de L'Écluse, director of the oldest botanical garden in Europe, the Hortus Botanicus in Leiden. Having introduced the bulbs to northern Europe from southeastern Europe,

L'Écluse had his valuable collection stolen and distributed around the continent. Although the tulip boom eventually collapsed like the South Sea Bubble, the theft of L'Écluse's collection is said to have founded the Dutch tulip industry.

WILD GARDENS

One 19th-century garden author who did as much for garden design and the fortunes of the bulb planter as his better-known colleague Gertrude Jekyll was William Robinson. The bachelor gardener died in 1935, leaving behind a 360-acre (146 ha) estate and an

TOOLS IN ACTION
Using a Bulb Planter

A flower bulb is a modified plant bud whose natural disposition is to start growing underground. Before planting, discard any damaged or diseased bulbs. Group bulbs in their possible planting places by the Edna Walling method, randomly tossing them onto the ground. (The Australian gardener Edna Walling, who died in 1973, was once asked how she wished to group some silver birch saplings: taking five potatoes from a bucket, she threw them all in the same general direction. "There!" she declared.) For heavy plantings, a two-handed, long-handled planter or turf plugger is ideal: place the cylinder, slightly open, in the position where the bulb is to go, press down on the cylinder with your foot, close the cylinder and withdraw the turf plug. Set the bulb or bulbs at the base of the hole, replace the soil plug and gently press it into place with the sole of your foot. Ensure the bulbs are placed at the appropriate depth and, on heavy wet soils, consider a layer of sand to improve drainage.

Elizabethan manor house, Gravetye in Sussex, and a pile of ash—he was an early advocate of cremation. He was also—despite having started out as a humble pot-washer at Curraghmore, County Waterford, Ireland—the richest and most influential garden writer of his time.

By 1859 Robinson had risen in the gardening ranks and was responsible for the hothouses on a country estate in County Laois. He is supposed to have abandoned his horticultural charges in the middle of the night following an argument with his boss, leaving the glasshouses open to the winds, and set off to find a new job in Dublin. It is an improbable tale, given his later eminence—he eventually secured a job with the

There are a variety of bulb planters available to the gardener from the basic flat trowel, above, to the long handled version that saves a lot of bending down.

Royal Botanic Society in London's Regent Park— but the anecdote suggests a telling characteristic: William Robinson did not suffer fools gladly. He did not care for topiary either (*see* "Hedge and Grass Shears," p. 121), and he positively abhorred fancy, or carpet, bedding, the practice of covering flowerbeds with swathes of garish perennials in carpet-style patterns.

Pelargoniums are commonly known as geraniums. They are a diverse group popularly used as bedding or houseplants in cool climates.

The craft of fancy bedding was the result of the mass import of new tender perennials such as *Verbena venosa* and little scarlet geraniums (more properly pelargoniums), coming into Europe and North America from South America and South Africa in the 1800s. Like the exotic birdlife of their native lands, these defiantly bright plants could cheer up the dullest of grey spring days. They were freshly raised from seed each year, or carried over winter in a warm glasshouse, and then planted in massed bands or swirling patterns around some mock-Greek urn or a pedestal topped with a rustic-looking basket.

Carpet bedding appealed to civic leaders, and their municipal gardeners vied with those of rival towns to produce floral coats of arms or beds of begonias that spelled out the town's name.

Stationmasters, meanwhile, competed with one another over their platform plantings, and in 1903 Edinburgh Parks unveiled a municipal clock entirely made out of flowers. In North America, too, fancy bedding grew in popularity through the 19th and early 20th centuries, with householders devoting the kind of resources to front-yard floral displays that their great-grandchildren would one day put into Christmas lights.

Robinson condemned it all—that "false and hideous art"—and advocated instead "the placing of perfectly hardy exotic plants under conditions where they will thrive without further care." His book *The Wild Garden* (1870) encouraged gardeners to plant bulbs such as daffodils and narcissi and, unusually for the time, alpine bulbs (his first book, *Gleanings from French Gardens*, 1868 led to *Alpine Flowers for Gardeners* in 1870). His ideas were soon persuading gardeners to rush out and buy a bulb planter.

Trowel

In 1845 one of Britain's oldest garden tool makers, Richard Timmins of Birmingham, offered a trowel as part of its Ladies' Garden Tool Set. It heralded a subtle change in attitudes to women gardeners—one that would be further helped along by the writing of one lady gardener, Jane Loudon.

Definition

A short-handled scoop-like device for ladling, cultivating and transplanting.

Origin

Named after the Roman *trulla*, a small ladle or dipper, the garden trowel was marketed as a ladies' garden implement in the 19th century.

The garden trowel exhibited every conceivable shape and form from the "strong-strapped trowel" (top left) to the elegant little fern trowel (bottom right).

E ven gardeners who bought nothing but the best were taken aback by the asking price of a second-hand, boxed set of hand-made tools that came on the market in 2012. The tools included two trowels, a fork, two pruning knives and a pair of snips or clippers. They were put up for sale at the London auction house of Christie's at between £2,500 and £3,500.

The tools were made of steel with handles formed from the hooves of muntjac deer. One trowel was decoratively engraved with a royal coat of arms, the other with the eagle of the French Second Empire. They had almost certainly never been used. The set had been manufactured by John Moseley and Sons of New Street, Covent Garden, London, in 1851 and placed that year on display at the first world's fair, The Great Exhibition at the Crystal Palace in London. They marked a turning point in the world of amateur gardening: the rise of the lady gardener.

The trowel is one of the few tools that have increased rather than diminished in variety in the last century or so. Contemporary trowels come in many forms, from long, hickory-handled items equipped with a carbon-steel blade to broad-shouldered stainless-steel trowels ideal for muscling about in the borders. There are slender lightweights and aluminium versions fitted with ergonomically designed rubber grips; Swiss-engineered stainless steel "dippers," the blade set at right angles in its plastic handle (brightly colored to prevent it being mislaid in the border); and, at the top of the range, the Jekyll, its polished steel scoop seamlessly riveted to a fine beech handle.

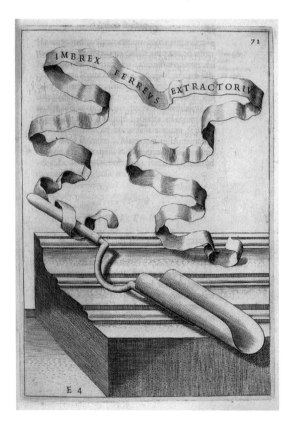

An engraving of a long-handled trowel taken from the Italian botanist Giovanni Baptista Ferrari's De Florum Cultura *published in 1633.*

Choosing a Trowel

There are trowels—and there are trowels. A cheap trowel will buckle above the blade, or else its handle will work loose after a night out under the stars. Garden magazines sometimes offer a free planting trowel as a special readers' gift; they are best avoided. The most versatile trowel will have a broad blade, occasionally marked with a useful depth gauge, and a tang and ferrule at the handle end. The handle itself should feel comfortable and solid. The best trowel will be versatile because it can be used for transplanting, weeding, ladling compost and even levering up some obstinate obstacle in the flowerbed (although manufacturers would prefer you did not).

THE FEMALE PROVINCE

In the late 1790s, when Anne MacVicar Grant was writing her *Memoirs of an American Lady*, only the most rudimentary of trowels were available. Anne was endeavoring to record daily life in Albany, New York, where "the care of plants, such as needed peculiar care or skill to rear them, was the female province." Respectable, well-off women—as she put it, "in very easy circumstances and absolutely gentle in form"— were considerably more fortunate than their impoverished sisters who were forced to earn a living as itinerant gardeners and farm weeders.

Anne and her women neighbors, nevertheless, lived in constrained times. In public, women were expected to be decorous, dignified and subservient. In the privacy of their own gardens, however, it was different. Anne and her gardens, where "no foot of man intruded after it was dug in the spring," were in the vanguard of a movement that would eventually lead to one (male) journalist at the *Strand Magazine* complaining of "the present day [1895] when ladies in bifurcated garments may be seen awheel in Piccadilly or enjoying a cigarette in the smoke room of their own club." He echoed the sentiments of the head of the Royal Botanic Gardens in London, Sir Joseph Hooker (1814–1879), who believed that gardening "as a life's work" for women was "almost an impossible thing."

Most young working girls were still destined for domestic service, and the same magazine in 1891 was recommending the local orphanage as the best place to recruit staff: "In these days, when good domestic servants are so difficult to get, the demand for foundling girls [orphans] is much greater than the supply."

Their employers, the mistress of the house, had enjoyed a long tradition of managing the household garden, but the 1800s saw them gain a new self-confidence in horticultural matters. A 23-year-old science fiction writer, Jane Webb, was about to hurry along the movement. In 1830 she married the 47-year-old Victorian garden writer John Claudius Loudon (*see* p. 16) and,

Colleges such as the Pennsylvania School of Horticulture advanced the cause of the woman horticulturist long before many other crafts and trades.

abandoning literary fiction (it was one of her novels that initially brought the couple together), devoted her energy to writing garden guides for women.

Jane was, at first, a little uncertain about the usefulness of the trowel. "A trowel is another instrument used in stirring the soil, but of course it can also be employed in boxes of earth in balconies," she offered in *Gardening for Ladies*. She was careful to appear submissive: "When I married Mr. Loudon, it is scarcely possible to imagine any person more completely ignorant than I was, of everything relating to plants and gardening." But she did her own research, never spoke down to her readers and, as a result, was widely read. When her *Gardening for Ladies* was published in 1840 it sold over 200,000 copies, including 1,300 copies on the day of publication.

The emerging self-assuredness of the woman gardener would lead to the establishment of ladies' horticultural colleges, often set up by wealthy, independent women themselves, such as Jane Bowne Haines's Pennsylvania School of Horticulture for Women, founded in 1910. In England 12 years earlier, the Studley College for Women had been set up by the countess (and mistress of the Prince of Wales) Daisy Greville to offer training and provide a place where "earnest-minded women" should not be expected to do household work but could "give their whole time to . . . become competent workers." Among its surplus women was Adela Pankhurst

(the daughter of the suffragette Emmeline Pankhurst), who gave up gardening to organize the Women's Peace Army in Melbourne, Australia.

These college founders, however, were anxious not to alarm their male counterparts. Frances Wolseley, founder of the College for Lady Gardeners at Glynde in Sussex, wished to "disarm any mistaken illusion which may have arisen that ladies wish to supplant men gardeners. . . . They do not wish to supplant able, clever men head gardeners, nor even to compete with them. They do desire however to assist as far as their strength allows, by lending intelligence, good taste, refinement, towards better cultivation of our great country." In her 1908 *Gardening for Women*, Lady Wolseley dealt with the trowel—"[It] should be chosen not too concave in the blade"—and advocated a small mason's trowel as a useful addition to the tool set. She could not have predicted the impact on horticulture of certain ladies with trowels such as Gertrude Jekyll, Vita Sackville-West, or Helena Rutherford Ely, who founded the Garden Club of America.

Pruning Shears

19th-century gardeners were almost superstitious about picking up a pair of French pruning shears and abandoning their pruning knives. In the expanding world of rose growing, however, Monsieur de Molleville's shears were making a big impact. They are still with us today.

Definition

A cutting device for cleanly severing hardy and half-hardy stems and stalks.

Origin

The origins of hand pruning shears are unknown, although they were introduced to the French garden around 1818.

B usiness interests in the West were trying to prise open the doors of trade to the world's most reluctant business partner, China, during the closing years of the 18th century. Still ruled successfully by the now ageing Qianlong emperor, China had begun to trade with the United States. New imported crops, such as corn and sweet potatoes, added to the staple diet of rice on which a rising Chinese population of 300 million depended.

In 1793, a hopeful King George III of Great Britain dispatched his ambassador, Lord Macartney, to parley with Chinese officials and try to negotiate a few new trade concessions. Macartney presented the emperor with £13,000 worth of expensive and curious gifts (they included the new Argand oil lamp, luxury Vulliamy clocks, a glittering planetarium—made in Germany—some charming musical automata and, quite probably, some Birmingham-made garden tools). Despite these and the precocious linguistic skills of an 11-year-old boy (the son of the mission's secretary, Sir George Staunton, provided the translation during the ceremonial meeting), the Qianlong emperor gave Macartney a frosty reception. As it was, the British monarch was told: "The Celestial Empire . . . does not value rare and precious things: nor do we have the slightest need of your manufactures."

Britain eventually broke down the trade barriers through other, more nefarious, means involving the opium trade, but for now she was left with no other option but to continue extracting through the port of Canton (now Guangzhou) limited amounts of tea, porcelain and silks. And roses: ever since the late 1700s, when one of the first of the pink-petalled China roses (*Rosa chinensis*) had found its way into Sweden with Pehr Osbeck—a pupil of the great taxonomist Carl Linnaeus—enterprising traders had sought more specimens of this rare, repeat-flowering rose.

A watercolor of a China rose, painted on rice paper by an anonymous Chinese artist, from the Royal Horticultural Society's Lindley Library collection.

CHINA ROSE

After the abortive Macartney mission, a group of traders stumbled on a plant nursery on the outskirts of Canton. It was called Fa Tee or "Flower Land." The Scottish plant hunter Robert Fortune described the place "where plants are cultivated for sale" on the banks of the Pearl River two or three miles (3–5 km) from the city when he was sent out by the Horticultural Society of London (later the Royal Horticultural Society) in the 1840s. "The plants," noted Fortune, "are principally kept in large pots arranged in rows along the sides of narrow paved walks, with the houses of the gardeners at the entrance through which the visitors pass to the gardens. There are about a dozen of these gardens . . . generally smaller than the smallest of our London nurseries."

The pot-grown roses caused a frisson of excitement when they reached Europe. There was even a Macartney rose (*R. bracteata*), which turned into a troublesome weed when let loose in the wilds of America. The European rose of the time was a mere midsummer bloomer, which broke briefly into flower for six weeks or so, depending on the weather, around June. ("Gather ye rosebuds while ye may," warned the 17th-century English poet Robert Herrick.)

Chinese roses, however, blossomed for longer and could, with careful pruning, be persuaded to flower a second time within the season. Their arrival led to a frenzy of cross-breeding just as a new French tool, pruning shears, was coming into use. While the more conservative gardeners continued to use their old-fashioned grafting knives, shears were usefully employed in grafting scions of the Chinese hybrid roses onto traditional, short-season roses such as the apothecary's rose (*R. gallica*)—a commercial rose grown for culinary as well as medicinal purposes—and the damask rose (*R. × damascena*), a favorite of the Middle Eastern and Indian perfumiers.

The new shears were perfect for managing the hybrid tea roses (so called because of their tea-like scent) and hybrid perpetuals such as the darkly sensuous "Général Jacqueminot," then all the rage in North America. Nurserymen had already produced the repeat-flowering and

Here were I and my servant on one side of some ravine, with our specimen boxes and other implements, gathering samples of everything we could find; there, on the top of the other, stood three or four hundred of the Chinese, of both sexes and all ages, looking down upon us with wonder painted on every countenance.

Robert Fortune, *Two Visits to the Tea Countries of China* (1853)

scented climbers, the Noisettes, by pairing the new English varieties such as "Parson's Pink" with the old musk rose (*R. moschata*) to produce the classic "Champneys Pink Cluster."

All this snipping, clipping and grafting ("Everybody cuts, but few prune," admonished the 17th-century Jean-Baptiste de La Quintinie) was diligently carried out by nurserymen and women alike, although John Claudius Loudon regarded shears as "particularly adapted for lady gardeners." He directed readers to the Sheffield manufacturers Steers & Wilkinson for "excellent instruments of this description" and explained in his *Encyclopaedia of Gardening* that "the French pruning-shears, by the curvature of the cutting blade, cut in a sort of medium way between the common crushing and pruning shears."

The new French style of shears, he reassured his readers, could deal with the most determined

TOOLS IN ACTION

Pruning

According to John Evelyn's book *The Compleat Gardener* (1693), the first vine pruning was carried out by a wild ass that broke into a vineyard and gnawed back some vine stems. When the vineyard owner noticed how quickly the damaged wood regenerated, he began the craft of pruning. Pruning shears need to be sharp to create a clean "draw" cut—crushed or mashed stems are susceptible to disease. Good-quality (and necessarily expensive) shears can be sent back to the factory for sharpening. However, drawing a fine abrasive such as a honing stone (nowadays not a stone at all, but a tool made of abrasive grains bound together with an adhesive) across the blades, taking care to follow the angle of the blade's bevel, will maintain a sharp edge to the cutters. Even the occasional wipe with an emery board will help.

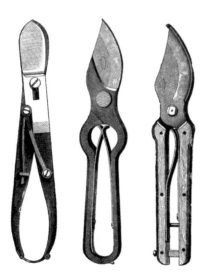

Nineteenth century pruning shears came in a variety of forms from ones equipped with sliding blades (far left) to the wooden handled, sprung pair on the right.

briar root, and English gardeners should set aside their disinclination to use them. "I know well the prejudice that exists in England among horticulturists against this kind of thing, and their almost superstitious regard for a good knife." (*See* p. 147.)

THE FRENCH CONNECTION

Loudon correctly described the origin of the pruning shears as Gallic, for the instrument had been developed by Count Antoine Bertrand de Molleville in the early 1800s. Molleville had been an unpopular governor of Brittany under his unfortunate and enfeebled king, Louis XVI. When Louis was taken from the Tuileries Gardens to the guillotine in the Place de la Concorde and beheaded by the Paris revolutionaries, Molleville, fearing for his own neck, fled to England. He returned to France after the Revolution and, perhaps musing on the efficiency of the machine advocated by Dr. Guillotin, came up with the shears' design. The tool was developed to cut in a variety of ways, including the guillotine-like action of a sharpened blade cutting against a flat metal surface, known as the "anvil secateur." Molleville's design, however, used two sharpened blades that clipped with a pincer-like action, an ideal device, he thought, for his fellow countrymen in the vineyards.

According to Loudon, French fruit growers were soon happily snipping away with upgraded versions: Loudon mentioned the Vauthier, with its wire-cutting notch on the back of the blade,

DEATH *of* LOUIS XVI. *King of* FRANCE. *who was beheaded Jan.* 21. 1793.

The inventor of the modern pruning shears, Antoine Bertrand de Molleville, escaped the guillotine, unlike his monarch Louis XVI, by fleeing France.

and the spring-assisted Lecointe shears. No one could have predicted that within a few years the vignerons would be throwing out their pruning shears as a devastating outbreak of phylloxera—the aphid-like vine pest introduced from phylloxera-resistant American vines in the 1850s—destroyed up to two-thirds of Europe's vineyards. While professional gardeners continue to debate the merits of their Swiss-made, Finnish-styled or Swedish-built shears, amateur gardeners are content to use anything resembling Count Molleville's original design.

Garden Basket

Despite all the modern materials available to the basketmaker, gardeners still prefer to pay up for a well-made, locally sourced basket that will hang comfortably from the elbow. Basketmaking ranks as one of the world's oldest handicrafts, and yet no two baskets are exactly the same.

Basketmaking began long before anyone started keeping records. Every culture has its own style of garden basket.

There are "bump-bottomed" baskets (so called because of their dimpled base) from the New York bushwhackers, and carrizo and totora reed baskets from the village women of San Martín and Cajamarca in Peru. Karelia in Finland is famous for its birchwood baskets, the Appalachians for their vegetable-dyed, split-oak carriers. At Katsuyama in Japan, lightweight vegetable baskets are woven from the local *madake* or bamboo, while in Botswana's Okavango Delta the women weave open baskets, designed to be carried on the head, from the leaves of the mokola palm tree. (Open baskets there are used for fish and vegetables, closed or lidded baskets for storing corn or brewing beer.)

The extraordinary range of garden baskets is a consequence of these vernacular traditions. Vernacular crafts (from the Latin *vernaculus*, a slave born in his or her master's house and therefore a native of the place) involve making do with whatever materials lie close at hand. Whether applied to domestic crafts or the home itself, the vernacular resulted in something that, as the English poet William Wordsworth wrote of the Cumbrian cottage, "may rather be said to have grown than to have been erected . . . so little is there in them of formality." And so the Mi'kmaq, a First Nation Nova Scotian people, developed a tradition for making basketry from

A universal device for carrying and carting garden produce, wood and wicker baskets represent one of the world's oldest crafts.

pounded oak splints. When the Black Loyalists, refugees from the American Civil War, arrived on the shores of Nova Scotia around the 1780s, they evolved their own distinctive basketry, working with reeds from the saltwater marshes and strips of beautiful red maple.

When the Shakers fled to America in the 1770s to enjoy, at last, comparative religious freedom, they set up communities that valued simple handicrafts such as that of hand-woven basketry.

STANDARD POTTLES

Nova Scotians and other seafaring people such as Basques, Bretons and the French in Normandy collected offcuts of sailcloth from the local sailmakers and turned them into cotton and canvas garden buckets. On England's watery Somerset levels it was willows and reeds that served the basketmaker, while London's Covent Garden Market was supplied by the traditional basketmakers of Fulham and Staines, where the willows, or osiers, lined the marshy edges of the Thames. Their standard "pottles"—small tapered baskets made to hold a half-pound (230 g) of strawberries—were specially made to fit into the larger "marne" carried on the heads of the Irish basketwomen who labored in the market and whose reputation for heavy drinking and creative cussing went before them.

Across the Irish Sea, W. B. Yeats wrote of meeting his love "down by the salley garden"— its name possibly derived from *sallow*, another name for willow. Families made a respectable living from their basketry on the shores of Lough Neagh, west of Belfast, and in Munster's Suir Valley. Here centuries of settled work were devoted not only to garden cribs and baskets, but to willow baby's rattles, dog muzzles, skibs (shallow wicker plates used to serve hot potatoes), and creels (carrying baskets made to suit anything from a donkey's pannier to a fisherman's pack). The invention of the cardboard box was as damaging to cottage industries as the

TOOLS IN ACTION

Care of Wicker Baskets

Anyone who has ever tried to break up an old, well-made wicker basket knows the resilience of wickerwork. Their three enemies are heat, damp and insect attacks. Avoid leaving baskets in strong sunlight or too close to the wood stove—they dry out and become brittle. And avoid allowing the basket to become wet. Clean baskets with a plain, dry paintbrush or wipe them down with a damp cloth. Soaking or submerging wicker in water causes the fibres to swell. If you are carrying produce from the vegetable garden, use a paper lining in the base of the basket to catch loose soil. Unless they are infested with woodworm, damaged baskets are not necessarily beyond repair; consult a basketmaker.

compulsory introduction of bicycle lights, which was said to have dealt a terrible blow to the wicker bicycle-basket trade, since handlebar-mounted baskets obscured lights mounted behind them.

The craft of garden basketry evolved a lexicon of regional names that are positively poetic. There is the wicker weave of the Catalan *coves* and *cargoleras* (the latter used for snails) and the West Virginian egg, rib, split and butterfly baskets, named either for their shape or their purpose. Britain has its Cumbrian swills and Worcestershire skuttles, Sussex spales or spelks, London pads (small rectangular baskets with a hinged lid designed to carry tender fruit such as peaches) and Wyre Forest whyskets and slops. Most renowned of all is the boat-shaped Sussex trug (*trog* was the old name for a boat), made famous, as we shall see, by one Mr. Thomas Smith.

A ROYAL BASKET

The shallow trug is designed to carry a basketful of blooms without crushing the ones beneath. Like the Cumbrian swills and Katsuyama baskets they are made from wood—although, while the Cumbrian "swiller" uses thin strips of oak, the Sussex trug maker employs strips of chestnut with ash for the handles.

The process of making a trug begins with a frame. This comprises the handle and the rim of the trug, and is fashioned from ash poles split into strips with a splitting axe. The strips are steamed and then bent on a jig, their ends "clinched" together with nails. The body of the trug is made from thin strips of chestnut, starting with a broad center board that is then flanked by overlapping side boards. Larger trugs are fitted with wooden feet.

Basketmakers such as the trug men used small, green (unseasoned) timbers, and they played a key role in the conservation of the local woodlands: since they relied on the regenerative quality of the trees, they were assiduous

protectors of the neighborhood's woods. Such a man was trug maker Thomas Smith. He followed his trade in the East Sussex village of Herstmonceux, but he had ambitions. In 1851 he took a gamble and paid to exhibit his trugs at the Great Exhibition at London's Crystal Palace. The Exhibition was designed to demonstrate the "Works of Industry of All Nations," and Mr. Smith was as proud of his traditional ware as any 19th-century engineer. Despite sharing space with Samuel Colt's firearms and the world's first iron-framed piano, Mr. Smith's baskets were spotted by an observant Queen Victoria as she toured the Palace, her beloved consort Albert by her side. Victoria promptly ordered a set for her extensive gardens.

The delighted trug maker was allowed to add the royal crest to his Sussex trug business, and when it came time to deliver his order he refused to trust either the coach or the Brighton train and carried them himself, in his handcart, the 60 miles (100 km) to London.

Sussex trug makers employ strips of chestnut with ash for the handles.

Soil-Test Kit

Clay or sand? Fertile or barren?
Rich or poor? The gardener's most
basic tool is the soil itself, and ever
since Roman times science struggled
to come up with a simple and reliable
means of analyzing it. Until, that is,
the soil-test equipment arrived.

Definition

A chemical test for measuring the
acidity or alkalinity of soil on a scale
of 1 to 14.

Origin

The implications of acid or alkaline
soils were finally recognized in the
early 1900s.

S oil is a growing medium made up of organic matter, rock and mineral particles, and its performance depends on the quantities and qualities of these three ingredients. Gardeners, like their plants, depend on it.

Soils range from the most fertile (a loamy brown crumble) to the most difficult (a chalky soil littered with flints). In between are clay soils (sticky, easily waterlogged, but high in nutrients), silt (fine-grained former river deposits), peat (black, spongy and acidic) and finally, sandy soil (light to cultivate and quick to warm up at the start of the growing season). A soil's fertility is also dependent on how acid or alkaline it is.

FERTILITY TESTS

As far as the Roman author Lucius Junius Moderatus Columella was concerned, the gardener needed only to check on weed growth to measure fertility. Columella committed to parchment the most lucid treatise on farming and horticulture produced during Roman times. His *De Re Rustica* (*On Agriculture*) ran to 12 volumes, and, while little is known about the author (he is thought to have been a Roman landowner born in Spain around 2,000 years ago), he had clearly studied the confusing business of soil fertility. He wisely looked for a bevy of weeds like rushes, reeds, grasses and even dwarf elder and briars when assessing a soil's potential productivity.

TOOLS IN ACTION

Acid or Alkaline?

Most garden soil-testing kits promise to be simple and quick to use. Most, however, rely on a degree of interpretation. (Sending a soil sample for a laboratory test will give more detailed results.) The standard test involves mixing a sample of soil with a chemical solution, then checking it against a color chart: yellow-orange indicates an acid soil, green is neutral, and dark green, alkaline. Plants generally do best with a pH of 6.5 to 7, which is the point where nutrients become most available to the plants. Acidic soils with a pH level below 6 can be improved by the addition of linden or ground lindenstone (avoid retesting within three months of applying linden); alkaline soils (above 7.5) need the addition of sulphur. Plants described as ericaceous, such as rhododendrons, do better in acid soils.

Columella proposed some simple fertility tests. The first involved dampening a sample of soil and kneading it; if it "sticks to the fingers . . . in the manner of pitch, it is fertile." A second test involved excavating a hole and then

The soil pH influences the final color of the flowers of mophead and lacecap cultivars of Hydrangea macrophylla.

Thomas Hill (or Didymus Mountain as he called himself), commended the practice of manuring the land to improve the soil in his Gardener's Labyrinth.

replacing the loose soil. If, after treading down, the soil more than filled the hole, Columella judged it to be good. A poor soil, he explained, would fail to refill the hole. The author even proposed a taste test in which the gardener took a tentative sip of a soil and water cocktail: a sweet tang meant a good soil, sour soil tasted just so.

Columella's soil analyzes were rediscovered during the early 1400s. His advice was set aside as a new wave of commentators entered the earthy arena, Thomas Tusser and Thomas Hill among them. Tusser, whose past is almost as obscure as Columella's (he seems to have failed as a farmer in East Anglia and died in poverty in London in 1580), commended the hop grower's soil in his *A Hundreth Good Pointes of Husbandrie* (1557). This was:

> . . . of the rottenest mould,
>
> Well dunged and wrought, as a garden-plot should.
>
> Not far from the water, but not overflown.
>
> This lesson, well noted, is meet to be known.

In *The Gardener's Labyrinth* (1577), Thomas Hill also recognized the value of good manure, recommending, for example, planting "lettice" seeds in a pellet of goat dung, a practice that guaranteed a plant "of a marvelous form and diverse in taste." Hill was a best-selling garden author. *The Gardener's Labyrinth* arrived at a time when the flower garden was starting to become established, and Elizabethan gardeners loved it (not least for some strange revelations: Hill claimed that the cucumber, which he oddly declared an "enemy of lust," was inclined to bend during thunderstorms.)

What, however, of the state of the soil? Or, as James Shirley Hibberd, the 19th-century writer, put it in *Profitable Gardening*: "What sort of stuff [have] you to work upon? Settle that first, and act accordingly." Gardeners had to wait almost another 50 years for science to provide them with a handy way of measuring a soil's potential. It was left to a Danish chemist, Søren Peder Sørensen, to reveal the concept of pH, the scale against which acidity or alkalinity can be measured. Working at the Carlsberg Laboratory in Copenhagen, he introduced the pH scale in 1909. Pure water, measured at a temperature of 25°C (77°F), has a pH of 7; any measurement below 7 is judged to be acidic, anything above, alkaline. Sørensen's discovery had all kinds of practical applications, but for the gardener, it meant a reliable measure of soil fertility that was eventually commercialized as the handy DIY soil test kit.

Dibber

The common garden dibber is hardly a headline tool. Over the centuries it was the reliability of the seed itself that preoccupied the gardener. Nevertheless, the dibber plays a key role in the garden by helping to sow and raise seeds.

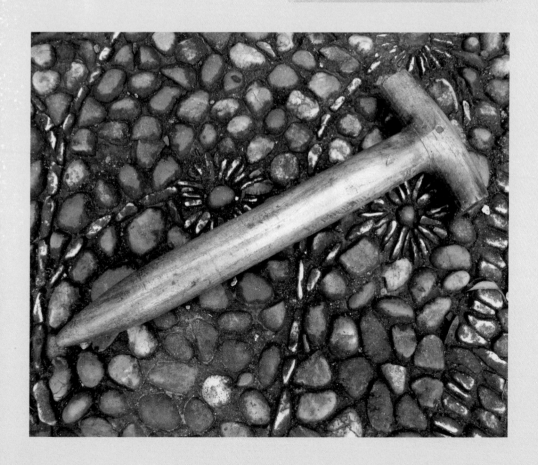

T he word "dibber" has a
modest ring to it, and
gardeners often make do without this
simple device for planting and sowing,
relying instead on an old spade handle
or a sharpened twig for such dibber-like
jobs as planting or pricking out seedlings.

Dibbers, or dibbles, are but one of a range of
helpful implements for the central business of
gardening: raising plants from seed. Such tools
include string lines and seed-spacing rulers,
heated seed starters, seed-drill hoes and
long-handled, trolley-like seed drills. There are
special survival seed safes designed to store seeds
during some undefined disaster: "Know that
your family will have food during a crisis," as one
website puts it. Then there are dibbers.

Even the inventive 19th-century manufacturers struggled to be innovative with this "short
cylinder of wood sometimes shod with iron," as
John Claudius Loudon described it. Yet they did
produce multi-pronged dibbers, which allowed
the gardener to drill several holes at once,
long-handled versions, and dibbers with a choice
of T-, J- and L-shaped handles. The new
materials of the 20th century saw the arrival of
elegant stainless-steel dibbers and an aluminum
trowel dibber that combines the attributes of
both tools.

All are designed to assist in that demanding
process of assisting the seed to transform into the
plant. The seed is the plant in an embryonic state,

*The metal-shod or plain wood dibber is
a device used to drill holes in the soil for
seeds or young plants.*

a ripened, fertilized ovule that ranges in
size from the dust-like orchid seed to the
heavyweight coconut.

SEEDS OF DOUBT

In the end it was not the dibber, but the science
of genetics that came to the gardeners' aid and
provided them with what promised to be the
perfect seed.

A product of agricultural science was the
dressed seed, wrapped in a coating of pesticide to
protect it against pests and diseases. Veteran
gardeners, meanwhile, were discouraged from
using old remedies such as rolling pea seeds in
paraffin, or dusting them with red lead, to deter
rodents. They were also, in the 1960s, discouraged from saving seeds as improved F1 hybrid
seeds, which had been developed in the 1920s,
began to emerge. The F of F1 stands for "filial,"
denoting the first generation son or daughter,
since the seeds were selectively bred from a pair
of parents. Like the genetically modified seeds
that would follow in its wake, this had one
advantage to the seed merchant and one
disadvantage to the gardener: the saved seed
would not set true, so the gardener was obliged
to return to the seed supplier the next season.
In time, however, many gardeners became

concerned about the loss of variety in the seed world. They were more inclined to follow the advice of a garden encyclopaedia from the 1930s: "It is a well-known fact that home-saved seeds give every satisfaction in the matter of germination and amateur gardeners should be encouraged to collect a certain amount of seed from their own gardens." This echoed the 16th-century sentiments of Thomas Tusser:

There is an artful simplicity to the dibber whether it was purpose made, fashioned from a bent piece of bleached wood or formed from a recycled fork handle.

> One seed for another to make an exchange
> With fellowly neighbors seemeth not strange.

As informal seed swaps gathered pace, non-governmental agencies invested resources in saving and preserving heritage seeds.

Arguments over selective seed breeding continue to rumble on. Genetically modified (GM) or transgenic seed, according to its producers, would not only increase food production but bring down growing costs by reducing the amount of money spent on herbicides and pesticides. Their opponents, who condemn these "seeds of doubt" for their hollow promises, counter that the new seeds are expensive and lock the buyer in to the seed supplier. In addition GM crops, they claim, would damage plant diversity, especially as farmers were forced to spend more on pesticides, tackling bugs that are resistant to the GM crop. The follow-up effect in the garden would be a loss of insect pollinators.

Activists mounted high-profile attacks on GM crops, but by the 1990s food from GM seeds was already on the supermarket shelf. Transgenic crops of cotton, corn, soy beans and rice, among others, spread from North America and Canada to India, South Africa, China, Brazil, Argentina and, after a moratorium, Europe. India, too, had resisted GM crops, approving only the use of GM cotton seeds, which were designed to disrupt attacks of bollworm, in 2002. Cotton production increased, as did the controlling interest of its seed industry, the fifth largest in the world, much of it dominated by the American GM seed producer Monsanto.

The domestic garden market has remained broadly resistant to GM seeds (although its gardeners almost unwittingly consume plenty of meat from beasts fed on GM meal). Meanwhile, non-governmental agencies such as the UK's Soil Association campaign against GM seeds and promote garden techniques such as the no-dig

The philanthropic John Ruskin would have approved of the utilitarian dibber.

method (*see* p. 14), which have brought the dibber back into the tool shed.

It was a tool that met with the full approval of the Victorian writer and gardener, John Ruskin. One of his first actions on taking over Brantwood, a country house in Cumbria, in 1871 was to replace the flowerbeds filled with blooms "pampered and bloated above their natural size"

with more natural plantings of flowers and trees. Deeply melancholic after his wife Effie eloped with the painter John Everett Millais, and highly philanthropic, Ruskin put £14,000 of his money into the St. George's Company, a charity founded to help disenchanted factory workers turn from their machines and find fulfilment in rustic labors. The Company would, he hoped, bring about social improvement and help the workers "gather their carefully cultivated fruit in one season" without recourse to "moving machines by fire." (Ruskin loathed the steam engine. He also hoped the Company's beneficiaries might make "scrupulous use of sugar-tongs instead of fingers.")

Working rural communities were established at Totley near Sheffield, Barmouth in Wales and Bewdley in Worcestershire, and, while the Company went on to become a charitable educational trust, initially these projects struggled to survive. The Barmouth project was only saved by the strange figure of the French exile Auguste Guyard, who strode around the village dressed in a grey cloak and a red fez, frightening small children and dispensing useful advice on protecting the vegetable gardens from salt spray—and, no doubt, how to make the best use of the strictly hand-powered dibber.

TOOLS IN ACTION
Using a Dibber

The choice of a dibber rests on the nature of the soil. Sandy soil will trickle back and refill the hole as the dibber is withdrawn. Here a tapered, stainless-steel design with a T-shaped handle allows the gardener to make the hole with one hand and release the seed or young plant as the dibber is withdrawn. A heavy clay soil will retain the shape of the hole, and a trowel dibber (dibber one end, trowel at the other) is useful. Open the hole with the dibber, and, having planted the seed or plant, use the trowel to lever the soil back over the hole. A seed-tray dibber equipped with a row of prongs is useful for multiple sowings. It can be purchased or made from a short handle of wood and dowel pegs.

Wellington Boot

The rubber boot swept away all rivals in the garden boot business when it went into mass production to cope with trench warfare during World War I. Certain gardeners and designers, however, refused to be parted from their well-worn and comfortable leather boots.

Definition

A waterproof rubber boot, designed to keep the feet dry—a gift to gardeners.

Origin

Rubber gardening boots (from the French *botte* and Spanish *bota*) were invented in the 19th century.

T he formidable Gertrude Jekyll, born in 1843, was destined to become one of the most influential garden designers of the 20th century. Though not so famous in her lifetime, she certainly was after her death in 1932. However, the lady who once declared that "no artificial planting can ever equal that of nature" was sufficiently celebrated to attract the attention of the Edwardian portrait painter Sir William Nicholson.

In 1920 he arrived to paint her portrait, but was kept waiting while she finished weeding one of her flower borders. Growing increasingly impatient, Sir William took out his sketchbook and drew her working boots. (What she was wearing at the time, we do not know.) *Miss Jekyll's Gardening Boots* was eventually finished in oils, and illustrated a pair of ankle-high, loose-laced black leather boots, with the sole of the right boot peeling away slightly from the upper. It was an intimate portrait that illustrated all the scuffs and scratches of age.

Fittingly, the picture was a gift for the 51-year-old architect Edwin Lutyens. Gertrude Jekyll knew him well. They met first when she was in her forties and he still in his late teens. (She called him Ned; he called her Bumps.) She was already sufficiently successful to be able to afford a new house, and more importantly a new garden, and she commissioned Lutyens to build it. The result, the 15-acre (6 ha) Munstead Wood in Surrey, was to become one of England's

TOOLS IN ACTION
The Boot Scraper

Many European gardeners, when watering their plants, endeavored to preserve their precious working boots by doing so barefoot. The arrival of the affordable "wellies," "bluchers" or "gumbies" was a blessing. The inexorable rise of the rubber boot also boosted production of the old-fashioned horseman's boot scraper. Boot scrapers range from sumptuously ornate 19th-century cast-iron collectibles to the strictly utilitarian H-shape set in a plain stone base. However, a basic boot scraper can be made from a tile of roofing slate set in a groove cut in a half-log and stood conveniently close to the back door.

No longer reserved for the muddy-booted horseman, the boot scraper was given a new lease of life by the rubber "wellie."

famous gardens. It also helped them both profit from their informal partnership.

Lutyens was a master proponent of the Arts and Crafts style. Jekyll's contribution was her eye for Englishness in the garden: a colony of pincushion-like London pride (*Saxifraga umbrosa* or *S. × urbium*) set in the cracks of a flagstone path; a silvery corridor of lamb's ear (*Stachys byzantina*), clipped lavender (*Lavandula* spp.) and yucca leading to a sandstone-colored Lutyens orangerie; a stone garden bench set against a sea of grey, green, white and silver flowers and grasses.

Jekyll, whose progressive myopia had frustrated her personal ambition to take up painting as a career, painted with plants instead. She spoke of "painting a landscape with living things" and creating garden scenes that worked "from all points and in all lights." She did so at Munstead Wood and on a string of commissions (she created over 400 gardens in Europe and America), and all in her familiar old boots. "I suppose no horse likes a new collar," she once said; "I am quite sure I do not like new boots." In the 1930s she was content with the pair she had bought when she had first met Lutyens in the 1880s. She was, in her choice of footwear as in her garden designs, flying in the face of garden fashion.

The "weeping wood" of Hevea brasiliensis *proved to have a transforming effect on what we wear in the garden.*

BEST BOOT FORWARD

While Jekyll was still a child, there had been a significant transatlantic exchange of ideas and patents involving rubber. Rubber, or what the Mayans and Aztecs called *cahuchu* (literally "weeping wood," which they collected from the rubber tree, *Hevea brasiliensis*), was in use long before the 18th-century chemist and clergyman Dr. Joseph Priestley discovered how these little balls of latex could erase or "rub out" pencil marks.

In the 1800s, American inventor Charles Goodyear and Englishman Thomas Hancock realized that raw rubber was magically transformed into a pliable, waterproof material when it was heated up with sulphur and lead oxide. The Scotsman Charles Macintosh had already patented his process of combining a layer of cloth with a layer of rubber to give the gardener the waterproof Mackintosh coat. It was not long before Hiram Hutchinson in France and Henry Lee Norris in Scotland were working to produce the most radical change in garden footwear since the Dutch first slipped on their *klompen*, the French their *sabots* and the English their clogs. Hiram Hutchison, an American industrialist, set up a shop making rubber boots in the little French town of Châlette-sur-Loing, south of Paris, in 1853. Henry Lee Norris, another American, founded the North British Rubber company in Scotland. This resulted in the French Aigle and Scottish Hunter boots, which, as the combatants lined up against one another in the trenches of World War I, were soon to do big business. (The growing sales of rubber tyres

Charles Goodyear's discovery that raw rubber could be transformed into a range of goods including waterproof boots did not meet with Gertrude Jekyll's approval.

would boost demand for rubber and result in the loss of Malaysian and Indonesian rainforests to vast rubber plantations.)

Had she chosen to, Jekyll might have ordered a pair of "Wellingtons," the boots, originally made of leather, that took their name from the first Duke of Wellington. In the mid-1800s the humorous magazine *Punch* conducted a review of the boots on offer and dismissed all but the Wellington, which it judged perfect, declaring: "[The Wellington] might be walked about in, not only as a protector of feet, but to the honour of the wearer."

It is no use asking me or anyone else how to dig, I mean sitting indoors. Better go and watch a man digging, and then take a spade and try to do it and go on trying till it comes and you gain the knack that is learnt with all tools, of doubling the power and halving the effort.

Gertrude Jekyll, *Wood and Garden* (1899)

Hat and Gloves

A pair of strong gloves and a sunproof hat may not spring to mind as one of the gardener's essential tools. However, commentators have been trying to convince the gardener of their necessity for several centuries.

Plants such as monkshood, euphorbia and these spiny thistles are best handled with a pair of garden gloves.

"Never perform any operation without gloves on your hands that you can do with gloves on," cautioned the German author Traugott Schwamstapper in his 1796 *Bemerken über die Gartenkunst* (*Remarks on the Art of Gardening*). Even weeding was "far more effectually and expeditiously performed by gloves, the forefingers and thumbs of which terminate in wedge-like thimbles of steel, kept sharp," advised Schwamstapper, who advocated "common gloves" for most other operations. "Thus, no gardener need have hands like bear's paws."

Schwamstapper was echoing the actions of Laertes in Homer's epic poem *The Odyssey*, written somewhere around the 8th century BC. While walking in his garden, Laertes wears a pair of gloves to stave off the brambles. Although some have forcefully debated the translation, claiming Laertes merely protected his hands with his robe, Laertes was sensibly protected against the thorns: he may even have been wearing the "rawhide glove" mentioned in Virgil's *Georgics*, written in the 1st century BC.

Britain's glovemakers were supplying leather gardening gloves as early as the 11th century. By 1340, the Pensions and Contributions account in the records of Norwich Priory record: "to the Scholars of Oxford, 2 shillings; to the cellarer for the cutting of herbs, 2 shillings; to the reapers of the Lord Prior, 6 pence; in gloves, 7 shillings."

Leather gloves were a stout defense against any briar—unlike the delicate chicken-skin gloves regarded as the height of fashion in the 17th century, or those made from the skins of unborn calves in Limerick, Ireland, their origins disguised with heavy scenting and much ornamentation. Such lightweights were no match for plants such as monkshood (*Aconitum* spp.), euphorbia, giant hogweed (*Heracleum mantegazzianum*), thistles or agave, which called for a pair of substantial, serviceable gloves.

When pineapples were grown in frames in early 19th-century Britain, leather-gloved gardeners were obliged to add hessian armbands to their protective armory when moving the spiky-leaved plants. Later it was the "native prickly plant . . . sometimes cultivated under the name of Cardoon of Tours" that necessitated "a strong leather dress, and thick gloves . . . to avoid personal injury," according to Scottish horticulturist Patrick Neill in the 1820s.

In 1880s Maryland, Mrs. Thomas Nelson, the first woman trustee of the Worcester County Horticultural Society, complained that young women had an aversion to gardening gloves.

"Young ladies don't like to don the garden hat and gloves," she declared, "because it is dirty work, and they are afraid they don't look as well as they would in fresh muslins, reading a novel, or doing some fancy work."

Best Against the Weather

A hat that has a dark color under the brim will absorb a little more reflected UV light. That brim ought to be 3 in. (80 mm) wide all the way around to protect your neck, chin, ears and cheeks.

Cheap gardening gloves will barely survive one season. Go for quality. Make sure the glove is a perfect fit; too big and they could slip off, too small and that could be the cause of cramp or aches. Either could result in blisters.

To test them out, put the gloves on both hands, clench your fists and mimic a range of movements that you regularly make while doing garden activities. If the gloves feel comfortable, if they don't slip or pinch, they're right for you. Select carefully: some types are suitable for wet jobs in particular, some for more delicate operations, and others are better when handling abrasive surfaces.

GET AHEAD, GET A HAT

The good Schwamstapper had also offered advice on a form of serviceable, all-weather headgear, suggesting that the gardener avail himself of "a broad-brimmed, light, silk hat, to serve at once as a parasol and umbrella." By the 19th century the hat, like gloves, were a reflection of status in English society, with the water-pot boy in a cap and the head gardener in a bowler hat—although attendance at the Royal Horticultural Society's Annual show at Chelsea required the full regalia of morning suit, white gloves and top hat.

The fashion-conscious lady gardener could, meanwhile, avail herself of the hat fan. The *Strand Magazine* of 1895 explained how the fan formed the peak of the cap, which, in rain or sunshine, could be opened to protect the lady. The hat possessed "the additional attraction of an accompanying curtain to shield the back hair."

A different solution was provided by an Australian rubber tire manufacturer, adapting their condom-making technique in 1964 for the mass production of disposable latex gloves. And as the hat and gloves became yet another of the gardener's important tools, Tom Oder, writing on the Mother Nature Network website (he had placed a hat as number 10 in his essential garden tools), related the story of the Georgia University horticultural professor who insisted that any of his students who came hatless to his outdoor lectures were required to write a special assignment—on skin cancer.

Garden Catalog

Garden brochures, filled with the allure of a bright and beautiful horticultural future, arrive in the dark days when nothing much stirs in the garden. Gardeners are inspired to order seeds and plants for the new season, and to forget those that failed to flourish last season.

Definition

An illustrated publication that provides information on seeds, plants and other horticultural paraphernalia.

Origin

"Catalog" derives from the Greek *katalogos*, meaning a comprehensive list. *Brochure* is French, and means something stitched together.

"It may be regarded as an indubitable fact, that all plants spring originally from seed," pronounced John Claudius Loudon in his *Encyclopaedia of Gardening*. His book was published in plain black and white in 1822. By the end of the century, color printing, although still in its infancy, was starting to transform not only garden books such as Loudon's, but that horticultural bible, the seed catalog. American seed growers were quick to catch on.

Having long relied on European imports, America's seed sellers began making inroads into the domestic market, led by men such as David Landreth—an Englishman from Northumbria who started a seed business in Montreal, Canada, before moving it to Philadelphia in 1784—and W. Atlee Burpee, who founded his famous seed company in 1876.

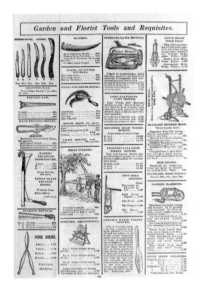

SPREADING THE WORD

As seed catalogs began rolling off the printing presses for delivery to distant homesteads by boat, coach, railroad and even Pony Express, the 800 or so U.S. seed merchants became relentlessly inventive. Along with their seeds there were special promotions and cash prizes for record-breaking crops. The 1888 Burpee catalog, for example, ran to 128 pages and promised "$25.00 and $10.00 for the two largest onions raised from seed purchased of us this year—the onions, or reliable affidavits of their weights, to be sent to us before November 1st." Then there were the persuasive testimonials: "The package of Vandergaw Cabbage you sent me did much better than the Large Late Flat Dutch."

Another American seed grower, Henry Field, began his business selling seed door to door in Shenandoah, Iowa, before joining with the Livingston Seed Company in Columbus, Ohio. Livingston was recovering from bankruptcy, advertising in the local newspapers and producing appealing-looking seed catalogs. Henry Field soon picked up enough ideas to leave and launch his own rival company under the slogan "Seeds That Yield are Sold by Field." By the 1920s Field had opened the first seeds grower's radio station over the shop in Shenandoah. Radio KFNF's call sign stood for "Keep Friendly! Never Frown!"

Simple seed lists gradually expanded into garden catalogs such as this 1906 treasure trove of horticultural paraphernalia (above).

Europe's seed merchants were restrained by comparison, but they were quick to capitalize on the business of color printing, none more so than Vilmorin-Andrieux, one of France's oldest seed suppliers. In 1782 the business was run from a shop on Paris's quai de la Mégisserie by a *Maîtresse Grainière* (Seed Mistress) and her husband Pierre d'Andrieux (he supplied seed to Louis XV's gardeners). The shop was handed on to their daughter and her husband, Monsieur Vilmorin. The Vilmorins proved to be astute marketers. They took on a team of 15 artists (including Elisa Honorine Champin, whose illustrations would eventually command high prices in the art world) to produce lavish and stylish watercolors of their plants and flowers. The images were used not only on their seed packets and in their catalogs, but published in a series of *Publications périodiques* alongside research into plant selection and heredity carried out by Louis de Vilmorin. By the late 19th century, Vilmorin-Andrieux had become the world's leading seed company.

In the days before seed catalogs, seed was simply sold in the marketplace or by the local plant nursery. London's largest in the 17th century, run by George London and Henry Wise, was said to occupy almost 100 acres (40 ha) of Brompton Park (later the site of the Royal Albert Hall and the South Kensington Museum), and was part of a lucrative business estimated to be worth over half a million pounds a year. Nurseries such as London and Wise shipped seed in from different parts of England, each of which had its particular speciality: Kent was renowned for its radishes, kidney beans, turnips, onions and "toker" or Sandwich beans; Berkshire for its cabbages and white-skinned, or Reading, onion seeds; Worcestershire and Warwick-shire for their white onions, asparagus, cucumbers and carrots. (According to London there was little seed to be had from Leicestershire since "the farmers, tho often rich, have seldom good gardens.")

In Reading in the early 1800s the seed merchants Suttons began distributing printed seed lists. Other nurseries had been advertising their wares in newspapers for half a century, but Suttons became one of the first to offer not only a price list, but handy planting

Different regions in Europe and the U.S. developed their own specialties when it came to sourcing garden seed.

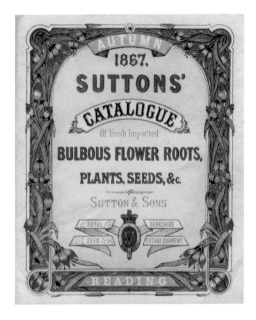

Sales soared at home and abroad when the railways arrived on the doorstep of seed companies such as the Reading-based Sutton & Sons.

advice as well. Suttons had started out selling seeds at Reading market in 1806. When the steam-powered Great Western Railway swept through town just over 30 years later, Suttons made the most of this new high-speed network, importing Dutch bulbs to sell to English gardeners and sending hundreds of thousands of colorful catalogs out with the mail van to Europe and Britain's distant colonies. "All flower seeds are sent free by Rail," promised the company, which by the 1880s was marketing its seeds "specially packed for India and the Colonies." Their efforts and those of rivals Webbs'—who shipped seed to Russia, Ceylon, New Zealand, Canada and even "Van Diemen's Land" (Tasmania)—ensured that the catalog would survive the growing season and act as a reminder to order

the company's seeds the following year. The 19th-century seed catalogs were, by now, expanding into every area of the garden, from wire rose temples to "mixtures of the finest grasses and clovers for Lawns, Croquet Grounds and Bowling Greens." Their breathless prose was not so very different from the promotional websites that would follow in their wake over a century later: "All our Lawn Seeds are carefully blended to proven specifications from only the highest quality seeds," promised one Melbourne website. Now as then, the garden catalog was packed with promise and hyperbole: "With great satisfaction we now introduce a new White Kidney which we have had under trial for several years. The yield has astonished experienced growers to whom we sent small parcels for trial," claimed Suttons of their Suttons Perfection potato in 1881. The brochure also carried glowing testimonials. As one immodest clergyman boasted: "I have cut Cucumbers from your Berks Champion three weeks earlier than any of my neighbors."

The seed catalog was aimed at those who paid the seed bill: the gentry, rather than the working gardeners, most of whom were illiterate. It prompted one Irish lady gardener to thank Suttons for "your beautiful and valuable Guide," which, she declared, had turned her own gardener—"a common laboring lad"—into "a most excellent practical Gardener. All his education has been from one of your Guides."

Garden Journal

Thomas Jefferson rose every morning to note in his daybook developments in the garden of his home, Monticello, in Virginia. Generations of gardeners ever since have maintained their own garden notebooks in order to chart what grew, what died and what the weather did.

Definition

A diary that documents horticultural successes and failures.

Origin

The garden journal or diary has a long and important history. Its name comes from the Latin *diurnalis*, "daily."

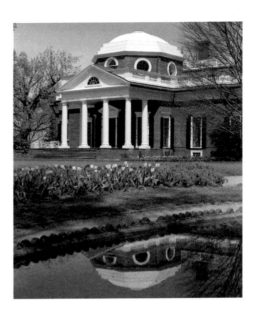

Developments on the estate and in the gardens at Monticello, Virginia, were recorded in a journal by Jefferson.

Thomas Jefferson was not only America's president between 1801 and 1809, he was also the architect of a Virginian masterpiece, Monticello. When he retired from running the country he devoted himself to noting developments on the estate. He would record the temperature first thing in the morning and at 4 o'clock every afternoon, together with the wind speed and direction and any precipitation. He noted the passage of migrant birds and new flowering plants, pencilling them in on the little ivory tabs that he carried in his pocket and transferring them to his garden book later as he "drudged" at his writing table.

Twenty years before Jefferson wrote his *Notes on the State of Virginia*, the scholar and naturalist Reverend Gilbert White, on the other side of the Atlantic, was scribing notes for his own *Garden Kalendar* in his best copperplate handwriting. "April 1761: 22. Hot burning weather which grew more and more vehement till the 25; and then a great deal of thunder, and lightning all night. The annuals are sadly scorch'd by the heat. May 8: Planted out leeks, savoys and two plots of endive."

Both Jefferson and White were following a long tradition of charting progress in a garden journal. The daybook, diary, journal or blog, call it what you will, is a non-essential but handy garden tool that details the gardener's successful seed sowings and abysmal failures, germination periods, plants that did well and those that died, good crops and disastrous ones, sources, samples and cunning new ideas and, above all, the vagaries of the weather. As one 21st-century blogger on easypeasyveg.net explained: "It's wonderful to look back over the years and see what you were up to . . . and what the weather was doing."

Old garden diaries also offer revealing insights into the quiet and otherwise private world of the garden. "This was a most exceedingly wet year, neither frost nor snow all the winter for more than six days in all. Cattle died every where of a murrain [a fatal infectious disease]," noted the diarist John Evelyn in 1648. His notes, published in 1818 as *Memoirs Illustrative of the Life and Writings of John Evelyn*, and those of fellow diarist Samuel Pepys, often shed light on the gardens of the time. 1641, for example, saw Evelyn in the monastery garden of St. Clara near Bois-le-Duc, where an ancient and overgrown pollarded lime tree had "issue[d] five upright and exceedingly tall suckers, or bolls." He had seen nothing like it before. Three years later Evelyn was struck by the "surprising object" of a fountain, "having about it a multitude of statues and basins" in the St. -Cloud garden of the archbishop of Paris; it was capable of "throwing water nearly forty feet [12 m] high."

Some garden journals tell us about the progress of other respected horticulturists, such as the Pennsylvanian Quaker John Bartram, who died in 1777. The Scottish plant collector David Douglas (1799–1834), traveling in Philadelphia in November 1823, noted in his North American journal: "We look round Mr. Dick's garden. I am again pleased to see *Maclura aurantiaca* [Osage orange]. The night's frost has made them drop their leaves, and the tender shoots were injured a little." Two days later he admired a cypress that had been planted by John Bartram in Philadelphia. Bartram had set up what became one of North America's first botanical collections on his farm at Kingsessing outside Philadelphia, but at the time of Douglas's visit the gardens were now tended by a Mrs. Carr. "Mr Carr to who she is married has but a moderate share of knowledge," gossips Douglas as he tells of the accidental felling of a famed oak, *Quercus × heterophylla*, which had distressed the great botanist. "Mr Bartram was not reconciled about [its loss] so long as he lived."

TOOLS IN ACTION
Contents of a Garden Journal

Gardeners often find it useful to record all or some of the following:

+ a plan of the garden
+ plant names, dates of acquisition and notes on performance
+ nursery information on care and protection
+ details on manuring, liming or fertilizing
+ vegetable crop rotations
+ sowing, planting and harvesting dates
+ quantity and quality of produce
+ plants, ideas and tools seen elsewhere
+ photos
+ regular weather reports

So much information requires a reasonably sized, hardback notebook. Some gardeners like to record factual information on one side and reserve the facing page for general comments. Increasingly digital, computer-based journals are being used.

By the time of his own death in 1818, the British garden designer Humphry Repton had made a particular use of garden notes. Repton was the successor to Lancelot "Capability" Brown, the former gardener's boy who was later hailed as England's greatest gardener after redesigning more than 170 parks and gardens. Brown was accustomed to undertaking the improvements to his clients' estates that he had himself suggested, and descriptions of Capability's grand garden designs spread by word of mouth among the aristocracy.

Repton's ideas, however, traveled farther afield, thanks to his Red Books. He used his journal notes to build up a plan of improvements (they included clever cut-outs and overlays), presenting them to the client in the form of a red, leather-bound book. He had produced over 400 of them by the time of his death, and their circulation among the landed gentry did much to spread the Repton look.

Humphry Repton was also persuaded to contribute sketches to a series of pocket garden "companions" produced by a bookbinder from Peckham in London named William Peacock. The clever Mr. Peacock perceived a market for an annual personalized leather-bound garden diary or "Polite Repository" that included a handy almanac as well as a fitted pencil for the discerning diarist to record his or her horticultural thoughts and observations. They fulfilled Peacock's promise of being "equally satisfactory in regard to general utility, as well as to ornament," and, beginning in 1790, sold for two decades. The personalized garden notebook or journal had arrived, and would continue to persuade the gardener to record the appearance of the first snowdrop and the onset of the first frost for the next 200 years.

For the American philosopher and writer Henry David Thoreau, the choice of pencil was as important as the diary itself. He preferred to use an A5-sized, marbled-covered notebook and a Thoreau pencil (the family business was in pencil making). The garden journals he kept between 1837 and 1861 grew into a writer's sourcebook when he came to write his *Walden; or, Life In The Woods*, philosophizing, for example, on the business of weeding the bean bed: "Consider the intimate and curious acquaintance one makes with various kinds of weeds . . . disturbing their delicate organizations so ruthlessly, and making such invidious distinctions with his hoe, leveling whole ranks of one species, and sedulously cultivating another. Many a lusty crest-waving Hector, that towered a whole foot above his crowding comrades, fell before my weapon and rolled in the dust."

The garden diary or "polite repository" became an essential aid in the gardener's world although, in the digital age, it faces a questionable future.

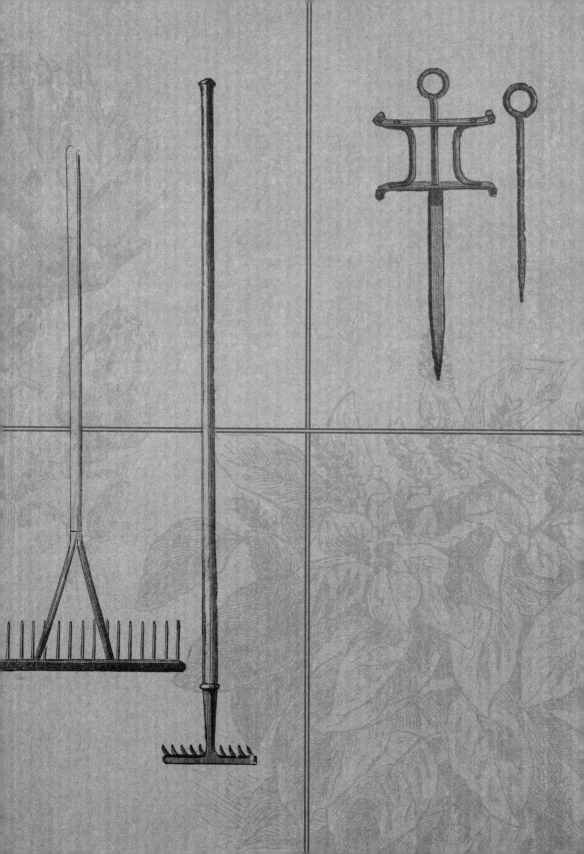

The Vegetable Garden

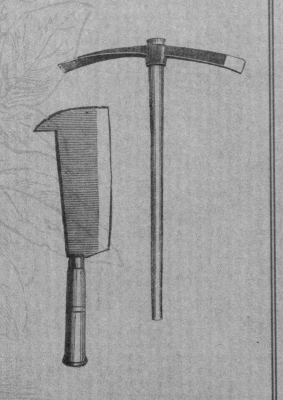

The vegetable patch arose out of the Anglo-Saxon *leac-garth*, the "backyard" taken by European migrants out into the new worlds. The colonists sent back an abundance of fresh, new vegetable varieties and came up with some new tools of their own.

Spade

An amiable old tool to the great Victorian garden designer and horticulturist Gertrude Jekyll, or a weapon "rusted with the blood of many a meadow" to the author and philosopher Henry David Thoreau, the spade has many faces and its history many twists and turns.

The Irish priest Fiacre was a fearsome fellow. Some time in the 600s he sailed across the sea from his native Ireland and, landing in Europe, headed for the Île de France to found a religious house. The wealthy Burgundian bishop of Meaux was persuaded to part with a plot of ground. But how much should he give? "Allow me all the land I can dig in a day," proposed Fiacre.

The following morning the Irishman took up his spade and began turning the chalky loam. By the time the sun had set, Fiacre had dug over 9 acres (3.6 ha) near to what is now St. Fiacre-en-Brie. A hermitage was founded on the site, and a shrine and chapel at Meaux. The priest (whose reputation as a gardener was eclipsed only by his reputation for relieving hemorrhoids) was beatified and his saint's day, 1 September, became the feast day of the patron saint of the spade. (His memory was also enshrined in the early horse-drawn cabs that plied their trade in Paris. Operating close by the Hotel St. Fiacre, they were known as *fiacres*.)

POOR MAN'S PLOW

Ireland's links with the spade, according to the 1930 *New Gresham Dictionary*, continued long after St. Fiacre. The spade, together with the pick and shovel, served an army of largely Irish "inland navigators," or navvies, who built the canals and railways that powered Europe's Industrial Revolution. "My God . . . how they work!" remarked a Frenchman watching them laboring on the line from Paris to Rouen in the 1840s. The navvies labored with the same tool as St. Fiacre, albeit much refined by Ulster's spade manufacturers.

The spade was the poor man's plow and its plain blade could be forged in any village blacksmith's shop, but as the Industrial Revolution introduced mass production and mass consumption, specialist spade factories arose to meet demand. Among them were the water-powered mills of Ulster, which from the late 1700s turned out garden spades with welded sockets in over 100 different designs. There were the narrow-bladed *loys* of south and west Ireland

The weapons . . . which should be handed down as heirlooms . . . are not the sword and the lance, but the bush-whack, the turf-cutter, the spade, and the bog-hoe, rusted with the blood of many a meadow.

Henry David Thoreau, *Walking* (1862)

Detail from Work *(1852–65) by Ford Madox Brown, showing navvies shovelling soil.*

and the dependable two-shouldered Ulster digging spade; mud and drain spades, peat-cutting *slanes* and trenching spades, the market for which rose inexorably during World War I. There were spades with double foot rests and others with the foot rest on the left or right, a custom that even became entangled in the complexities of Irish politics when it could be whispered ominously of a Protestant or a Catholic: "He digs with the wrong foot."

In the 1950s and 1960s, the mass production of garden tools in left traditional spade makers fighting a losing battle. One by one the specialists were driven out of business by cheaper imports. The loss of the traditional hand-made spade, however, created a premium market for old, second-hand spades made from welded steel and fitted with fine ash or hickory handles.

Digging the ground with a spade was always regarded as an essential operation. When the impoverished clothier Gerard Winstanley was ruined during the English Civil War in the 1640s he formed a company of "diggers" who took their spades to Cobham Heath, a piece of common land in Surrey. (Common land, despite its name, was held in private ownership but let to local commoners for grazing and other rights.) "In Cobham on the little heath the digging still goes on and all our friends, they live in love as if they were but one," wrote the revolutionary digger, even as the authorities sent in the soldiers to pull down their beans and uproot their parsnips. (The ubiquitous potato had yet to make an impact on the gardens of England.)

TOOLS IN ACTION

Digging a Planting Hole

Around the time that the Romans invaded Britain, the gardener Lucius Columella was writing his horticultural instructions on how best to dig a hole. Planting holes, dug in poor ground and filled with fertile soil, were a popular method of overcoming adverse soil conditions, especially in the arid Mediterranean. Columella advised taking the spade to the ground a year in advance. The hole, he wrote, should be shaped like an oven, narrow at the top and wider at the bottom so that plant roots could spread and be protected from dry or cold conditions. To make the soil more friable he advocated burning straw in the pit before planting.

DIGGING FOR VICTORY

The necessity of digging over new ground before planting was central to another campaign three centuries later, in World War II. As supplies of nitrogen were diverted away from agriculture and into explosives, the British mounted a campaign to Dig for Victory.

"Half a million more allotments properly worked will provide potatoes and vegetables that will feed another million adults for eight months of the year," promised the British agriculture minister at the time. "Let Dig for Victory be the matter for everyone with a garden or allotment." In 1941, as Eleanour Sinclair Rohde started work on *The Wartime Vegetable Garden* and the Royal Horticultural Society published its *Vegetable Garden Displayed*, patriotic Britons grew a record 1.3 million tons of fresh food. Translated into German at the war's end, the work was used to help in the reconstruction of Europe's vegetable gardens.

The United States, which donated 90 tons of vegetable seed to the British war effort, mounted its own Dig for Victory campaign, despite the opposition of the U.S. Department of Agriculture. At the time national food surpluses were at an all-time high and the USDA was resistant to any plan to "plow up the parks." Nevertheless, the Burpee Seed Company, in response to patriotic appeals from American gardeners, launched its Victory Garden Seed Packet in 1942 and an estimated 4 million Americans took their

The famous image of the left foot on a spade with double foot rests was designed to inspire the wartime, amateur gardener to grow more vegetables.

spades to turn over their own backyards. Even President Franklin Roosevelt ordered the White House lawn to be dug up and planted with cabbages, beans, carrots and tomatoes.

Hoe

No tool has appeared in such divergent forms or been used for such a variety of purposes as the hoe. While legend credits the mythical Chinese ruler Shen-nung the Divine Farmer with its invention, the hoe has made its mark on virtually every culture across the globe.

Definition

A long- or short-handled implement for breaking up and loosening the soil and weeding.

Origin

The name traveled into Middle English as *howe* from the French *houe*, related also to the Germanic *houwan*, "to hew."

As long ago as the third millennium BC, a Sumerian creation myth invoked a god with a hoe: Enlil. Enlil was said to have first created daylight with his golden hoe, a colossal implement fitted with a blade of lapis lazuli, before creating mankind with his hoe and a brick mould. Around 1770 BC, references to the hoe appear in the Babylonian *Code of Hammurabi*, one of the oldest social charters in existence. It was mentioned again in the 8th and 7th centuries B.C. in the books of the prophets Samuel and Isaiah.

Hoeing tools advanced in the early 13th century A.D. with the introduction of iron and steel blades.

From earliest times the hoe kept pace with technology, finding its form in copper and bronze until, around the 13th century, iron and steel blades so improved its effectiveness that it spread across the world. The English scholar Anthony Fitzherbert described using a hoe in his *Year Book* of 1534. The gardener takes up his "wedynge-hoke" (weeding hook). "In his other hand he hath a forked stycke a yard [90 cm] longe, and with his forked stycke he putteth the weed from him, and he putteth the hoke beyond the root of the wede, and putteth it to him, and cutteth the wede fast by the earth."

The hoe was a valuable and prized possession and apparently worth stealing. In 1763 the *Pennsylvania Gazette* reported how Adam Reed of Lancaster County was escorted by law-enforcement officers to the place where he had "hid in the ground" two spades, four shovels and four "grub hoes." During the Revolutionary War,

as hostile forces neared Philadelphia, John Jones of Southwark deposited his tools, including five garden hoes and one grub hoe, for safekeeping with a Captain Christian Grover. When in 1778 the tools disappeared, Jones offered a generous reward for their return.

Tools such as these were custom-made and difficult to acquire. They were manufactured by local craftsmen such as those at Philadelphia's Sign of the Scythe and Sickle where, in the 18th century, cutler Thomas Goucher and blacksmith Evan Truman offered "all kinds of edge tools and all sorts of hoes." The American poet, naturalist and philosopher Henry David Thoreau, certainly valued his hoe. In *Walden; or, Life in the Woods* (1854), he writes of taking a break from his work and, pausing to lean on his hoe, observes the world around him. "When my hoe tinkled against the stones, that music echoed to the woods and sky, and was an accompaniment to my labor which yielded an instant and immeasurable crop."

Thoreau was a dedicated abolitionist, vehemently opposed to slavery in the U.S. which had its roots in the slave trade with Africa. Here too was a continent wedded to the hoe, a place where people not only composed songs of praise to the implement, but even used it as a form of currency. The hoe was a recurring image in the tribal songs of northwest Tanzania, where, as in so many cultures around the globe, the tool symbolized long toil and hard work. For the iron-smelting Batembuzi "blacksmith kings" of Bunyoro, one of East Africa's most powerful kingdoms, the hoe served as their tribal insignia. It was customary for a tribesman, standing as a new community leader, to symbolically upend his hoe. The first hoe produced from each smelt was traditionally gifted to the chief by way of tribute.

In some regions a chief would pound out a rhythm on an old hoe to set the growing season in motion, while the home of his first wife (sometimes titled "queen regnant of the hoe") carried on its roof an emblematic vessel filled with earth in which an iron hoe was planted.

TOOLS IN ACTION
Using a Hoe

The length of a hoe should be in proportion to your height, with a handle long enough to allow you to work in an upright position, preventing possible back problems. Depending on the type of hoe, use it for loosening soil in order to aerate, for weeding, chopping roots, creating furrows in which to plant seeds, hilling (piling up soil around the base of a plant) or leveling and edging. When working with a Dutch hoe, maintain a flowing, sweeping movement. The blade should brush the surface and drop beneath the earth without hindrance. Standard, commonsense maintenance rules apply: keep the blade sharp and clean it after use.

By the 19th century the hoe was being traded for livestock and grain, one hoe being worth two goats. "The hoe is prosperity. It will bring you cattle," sang the villagers as the Bena people in Njombe, southern Tanzania, put the asking price for a bride at three iron hoes.

I kept Homer's *Iliad* on my table through the summer, though I looked at his page only now and then. Incessant labor with my hands, at first, for I had my house to finish and my beans to hoe at the same time, made more study impossible.

Henry David Thoreau, *Walden* (1854)

Triangular hoe

Half-moon hoe

Cast steel hoe

Canterbury hoe

Dutch hoe

Know Your Hoe

The hoe comes in a bewildering variety of shapes and sizes. In *The Ladies' Companion to the Flower Garden* (1865), Jane Loudon and Charles Edmonds dealt with this problem by reducing the hoe family "to two classes, the draw hoes, which have broad blades, and are used for drawing up the earth to the roots of plants, being pulled to the operator; and the thrust or Dutch hoes, which are principally used for loosening the ground and destroying the weeds, and which the operator pushes from him."

The basic shape of the draw hoe has not changed since Neolithic times. It was and still is a groundbreaking tool, harking back in form and mode of use to the mattock (*see* p. 65). In German the verb *hacken*, "to hoe," also means to hack, to chop, to peck at, while the noun *Hacke* denotes both a pickaxe and a hoe—or in Austria, a hatchet.

The farm laborer's favorite, used the world over, is the classic eye or grub hoe with a handle that fits through the "eye" in the blade. The grub hoe is featured in Jean-François Millet's painting *L'Homme à la houe* (1860), the picture that inspired the American Edwin Markham to write his famous poem similarly titled "The Man With the Hoe."

Almost every task had its special hoe from the triangular or eye hoe, left, to the classic Dutch hoe, right, designed to slice away weeds just below the soil surface.

The grub hoe is a variant of the tool also known in various regions as the farmer's grab, dego or pattern hoe, useful for turning soil, weeding and thinning (removing surplus seedlings from the vegetable row). A variation of this is the wide-bladed and three-tined cultivating eye hoe, used for harvesting and turning over the loam. If one side of the hoe's head is furnished with tines or spikes, or if the blade is divided, it can be operated like a fork, cultivator or rake.

One 19th-century French garden catalog produced by Truffaut advertised its *serfouette à oignons*—a combined "hoe fork" used specifically for onions—at 1 franc 20 centimes. The French take their hoes seriously: there are said to be more names for hoes—*hoyau, bêchoir, féchou, écobue, besoche, bêchard, essade, déchaussoir, moutardelle*—than there are actual tools. Many of these implements are adapted to the specific needs of their region and have earned a particular local name. Among bifurcated or divided hoes are the *marre* from the Médoc, the Provençal *bigorne* and the *mègle* from Burgundy, all examples of the French passion for inventing implements while at the same time preserving a tradition.

known as a *schoffel* (or shovel); it is also referred to as a "scuffel hoe" by the English. (The French, however, claimed it as their own, describing it as the *ratissoire normande*, or Normandy hoe.) Some versions were produced with a sharpened edge on the back of the blade as well as the front, allowing for weed clearing on both the push and the pull strokes. Yet, according to the Dutch toolmaker Jaap Sneeboer of Sneeboer Manufacturing, the Dutch hoe was not a tool of the Netherlands. (He nevertheless believed it to be the best-functioning design of all, principally because the end of the blade was visible to the user and the earth flowed over this blade.)

Finally, there are the scuffle or stirrup hoes formed, as their names imply, so that they can be pushed and pulled through easily worked, friable soils (the similar hula hoe is equipped with a swiveling head), and the Paxton hoe, a Scottish innovation ideally suited for use in confined spaces.

Making light of the hoe's longstanding association with sustained toil and hard work, the 19th-century dramatist and wit Douglas William Jerrold observed in *A Land of Plenty*: "Earth is here so kind, that just tickle her with a hoe and she laughs with a harvest."

In Australia, meanwhile, the three-pronged hoe is a traditional gardening tool, useful for displacing small weeds and aerating the soil. Very different is the Warren or pointed hoe, with a blade shaped like an arrowhead. Originally designed in North America, it is suitable for digging holes, cultivating between plant rows, and opening and filling seed drills. The *Encyclopædia Britannica* noted that the crane-neck or the swan-neck hoe "has a long curved neck to attach the blade to the handle; the soil falls back over this, blocking is thus avoided and a longer stroke obtained."

The first recorded use of the Dutch, or thrust, hoe was around 1750. As a traditional tool from West Friesland it was originally

Mattock

Since it fell out of favor in the West, the mattock is more likely to be seen working the ground in a Botswana paddock than a Boston backyard. And yet this is a highly versatile hand tool, useful in breaking new ground or excavating old roots.

The mattock, like its close cousin the pick axe, is one of the oldest tools known to man.

Old mattocks lurk neglected in the dusty corners of tool sheds. Their utilitarianism has been superseded by ranks of powered cutters, cultivators and root pullers. Yet this is the most venerable tool in the garden.

The mattock, or in Ireland the *matóg*, was the forerunner of the spade and combined the attributes of hoe, trencher and root-digger. It was a close cousin of the adze, the wood-butcher's fashioning tool, and the poleaxe, used to fell a farm beast with a quick blow to the skull. It evolved into a variety of specialist tools including the pickaxe and the miner's mandrel, and even inspired the pulaski, a special wildfire-fighting tool invented by U.S. firefighter Ed Pulaski following the outbreak of bush fires in Idaho in 1910.

Essentially a cultivating tool, the mattock was used to break new ground and grub out the undergrowth when establishing a garden.

Let servant be ready with mattock in hand
To stub out the bushes that noieth
[encumber] the land.

recommended Thomas Tusser in 1557. The mattock might serve to lay down a drain, excavate an irrigation trench or clear out a ditch. The mattock was making its mark on the landscape long before Mr. Tusser commended its usefulness. It was used 12,000 years ago, when gardeners raised their first wheat, barley, lentils and chickpeas in the Middle East; 8,000 years ago, as the Chinese gardeners first sowed their rice and millet; and 7,000 years ago, as Mediterranean gardeners planted the first almonds, grapes, lettuce and olives. Around that time the earliest corn and cotton crops were planted in Mexican soil, cleared with mattocks. Six thousand years ago, Peruvians had mattocked the ground out ready for the first crops of sweet potatoes, chilies, potatoes, avocados and groundnuts. Five thousand years ago it was the turn of the North Americans to heft the earth with wood and stone mattocks in preparation for the first harvests of goose grass, sunflower and sump weed; and, in central Africa, mattocks prepared the way for the first millet sorghum, yams, oil palms and coffee.

The mattock was as useful at prizing big boulders out of the earth as it was at breaking up virgin ground.

Using a Mattock

The mattock is more useful than it first appears, especially in preparing new ground, excavating the roots of old shrubs and bushes, or moving large plants from one part of the garden to another. The metal head is attached to its wooden handle through the eye hole in the head. Since the mattock is handled like an axe, swung over the head and buried with full force into the ground, it is vital that the handle is sound. Check for rot or woodworm, renewing the handle if it is at all compromised. The blade of the mattock can be honed with a sharpening stone or a metal file, but since it is the brute force of the tool that makes the most impact, this is rarely necessary. It is also a tool that is best handled without gloves (to avoid slippage): a well-oiled handle should prevent blisters.

THE MATTOCK UNEARTHED

Occasionally the remains of some ancient mattock, once held in the hands of these pioneer gardeners, rises to the surface. One mattock head, formed from the antler beam of a red deer and found on the Thames riverbank, was deposited with the British Museum in London. Conservators estimated that it had previously carried a pole handle, wedged in a neatly drilled hole, at least 4,200 years earlier. Another, older

still and found near Kew Bridge, again on the Thames, would once have ridden into the field on the shoulder of some Neolithic man or woman at the very dawn of farming. A small bronze mattock head unearthed in California and passed to the San Jose Rosicrucian Egyptian Museum was probably turning the ground only a little later than those mentioned by the Greek poet Hesiod. In his epic 800-verse farming almanac, written around 700 BC, he advocated sowing seed and letting a slave "follow a little behind with a mattock and make trouble for the birds by hiding the seed."

According to John Claudius Loudon, 19th-century Greek gardeners were still "unacquainted with the spade, and only use a mattock for turning [the soil]." The ground, he judged, was "in general . . . ill prepared." After the Greeks came the Romans and, according to Loudon, "the marra, a hoe mattock." Such a tool turned up as part of a hoard of Roman items at Lakenheath, England, in 1985. The mattock gave rise to the pickaxe; both tools, along with the shovel, pioneered the canal- and railway-building era until the advent of the steam shovel. Although a small version evolved for working the ground at close quarters, the mattock mostly fell out of use. Yet, as Hesiod promised, using the mattock almost guaranteed a fine crop of corn where the ears would "bow to the ground with fullness . . . and you will sweep the cobwebs from your bins. And you will be glad."

String Line

Where would any gardener be
without a ball of string? Jute, the most
important and versatile crop after
cotton, was the plant of choice when
it came to supplying garden twine.
However, while it beat flax, natural,
biodegradable jute was affected by
the arrival of plastic string.

T wine, or cord, is an essential tool. String reels, or row markers, keep the vegetable patch in order. Ties secure plants in place and hold together bunches of home-harvested asparagus spears, rhubarb or runner beans. French family kitchens are hung with strings of dried mushrooms in the autumn, while garlands of red Basque peppers are strung out to dry on the whitewashed walls of Pyrenean farmhouses.

No mistress could manage her herb garden without a generous supply of cords to tie up the bunches of lavender, thyme, rue and sage that were hung from the rafters of the outbuildings as the nights drew in. In days gone by, the mistress's herb patch was the most important part of the garden, more vital even than the orchard since her versatile plants played not only a culinary role, but a medicinal one too. In his 1597 *Herbal*, the Elizabethan botanist John Gerard heralded them as treasures: "And treasure I may well term them seeing both King and Princes have esteemed them as Jewels." Herbals such as Gerard's codified all plants, not just those we think of as herbs today, listing their essential qualities for the benefit of both the apothecary and the housewife. So the rose, which "doth deserve the chief and prime place among floures," not only strengthened the heart and gave a delectable taste to cakes and sauces—it also, according to Gerard, eased eye pains.

The cords that tied together the bunches of herbs in Gerard's own garden at Holborn,

A sharp knife and a ball of twine were among the essential items carried by 16th century botanists John Gerard, author of this Elizabethan herbal.

London, could be made from virtually any vegetable fiber. There was "corn straw", left over after grain had been harvested; or the stringy cellulose fiber, coir, that surrounded the coconut. Earlier still, there were water reeds in ancient Egypt and hemp fibers in China and other parts of Asia; there was cotton (*Gossypium* spp.); there were henequen and sisal taken from fibrous American agaves (*Agave fourcroydes* and *A. sisalana*); and flax (*Linum usitatissimum*), the source of linseed oil from Egypt and India.

New Zealand "flax," described by the explorer
and cartographer Captain Cook, was not a flax
but a *Phormium*; the error is understandable
since, he wrote, "of the leaves of this plant they
[the Māori] make also their string, lines and
cordage." For flax was one of the oldest sources of
string in Europe. Gerard knew it well: "Flaxe is
sowne in the spring, it floureth in June and July."
It was processed before it was "taken up and
dried in the Sun, and after used as most huswives
can tell better than my selfe," Gerard tartly
remarked. Painters and picture makers made use
of flax-seed oil and, when it was "stamped with
the roots of wild Cucumbers," flax could draw
out splinters and thorns and mend broken bones.

To process flax, the plant was pulled up and
then left to dry in blond, "flaxen" streamers on
the ground. Later it was taken for retting, that is,
soaked for weeks or months to soften the fibers
(the smell of retting flax was a cause of much
complaint) before being rolled up when dry and
taken to the flax mill. The preparation of flax was
essentially a cottage industry, but growing
demand drove 18th-century entrepreneurs such
as Shrewsbury's Benjamin Benyon to devise a
means to mechanize production. He finally
succeeded in the 1790s and set up a flax thread
mill in his home town, an enterprise that was
said to have provided his business backers with
"ample fortunes" for the time.

During the 19th-century gardening boom,
every bank clerk, butcher and baker aspired to

Rippling the Flax heads, pages 50—51.

Threshing the Flax, pages 50—51.

*There were "ample fortunes" to be made from
mechanizing the cottage craft of flax spinning.*

his own backyard garden. As the number of
gardens and allotments increased, so did the
rows of scarlet runner beans and decorative
dahlias, all demanding their piece of string. The
market was met in part by the raffia trade. The
raffia palms of the African mainland, Madagascar
and South America produced lengthy fronds—
the largest in the world at over 80 ft. (25 m)
long—which were, and still are, sold as a
rudimentary and decorative garden string.
However, it was jute and not flax that was to
dominate the market.

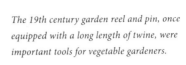

The 19th century garden reel and pin, once equipped with a long length of twine, were important tools for vegetable gardeners.

THE JOY OF JUTE

Jute was a vegetable fiber that had been used for centuries in what is now Bangladesh. The East India Company, Britain's exclusive trading arm in India for 200 years, began importing it to the mother country. Jute was a versatile material, second only to cotton. It could be twined into string or woven into the coarse cloth, hessian, that was used for horticultural sacking. It could be woven into the soles of the Spanish gardener's espadrilles. When it was discovered that raw jute treated with whale oil could be easily spun into a fabric, the whaling fleet and the "brabeners" (cloth weavers) of Dundee, Scotland, worked together. It was a relief for the whalers, whose business had begun to decline, and by the 1860s the business had created a bunch of wealthy jute barons.

Then along came George Acland. He was a former East India man who understood more about jute than flax, and had been one of those who persuaded the Dundee weavers to switch to jute in the first place. In 1855, however, he took some of the business back to British India, to Rishra on the banks of the River Hooghly near Kolkata (Calcutta). British interests dominated southern Kolkata, and Acland's jute mill eventually began exporting the finished product to Britain and other parts of Europe, undercutting the Dundee manufacturers. By the 1890s the banks of the Hooghly were lined with new jute mills and Dundee's jute barons were all but broke.

Not until the middle of the 20th century did jute face a new rival: plastic string. Plastic string was brightly colored, difficult to break and did not rot in the glasshouse. By the 1980s Bangladeshi farmers were burning their jute harvests in protest at their losses. Even the Adamjee jute mill at Narayanganj, once the world's largest, struggled to survive. It finally closed in 2002, a few years before the beginning of a market revival in natural fibers.

TOOLS IN ACTION

How to Tie Plants Without Chafing

Jute string is a useful and natural material to use as a plant tie. A tie, however, can cause more problems than it solves if it is made so tight that it bruises the growing plant stem, or so loose that it garrottes the plant in a strong wind. Soft-stemmed plants such as tomatoes, string beans, dahlias and chrysanthemums are best tied to a cane with a figure of eight: once around the support, and once around the plant stem. Harder plant stems such as raspberry canes can be double-tied, with the string looped twice around the stem and its support.

Billhook

As closely related to the pruning knife
as it is to the axe, the billhook had,
and still has, a variety of uses around
the garden, from trimming brushwood
to pruning vines. In Europe the billhook
has a long history as a hedge maker
and, with a growing interest in the old
craft, it could yet return to favor.

Definition
A venerable old tool, the billhook
is designed for heavy pruning,
coppicing and hedge laying.

Origin
The billhook was as familiar to
the Romans as to the 20th-century
hedge layer.

G ardening at close quarters can be a problem, as Mrs. Nickleby noted in Charles Dickens's *Nicholas Nickleby* (1839): "The bottom of [the neighbor's] garden joins the bottom of ours, and of course I had several times seen him, sitting among the scarlet-beans in his little arbour, or working at his little hotbeds. I used to think he stared rather, but . . . as we were newcomers . . . he might be curious to see what we were like." Suburban gardening protocol can be a mystery: do you greet your neighbor, or pretend that they are invisible? The sensible solution was a fence. *Gode hegn gøre gode naboer*—"Good fences make good neighbors," as the Danish gardener put it.

The provision of effective physical boundaries has exercised gardeners for centuries. While the often feeble fence panel serves many, the laid hedge is still a sensible and long-lasting solution. And the tool most likely to be employed in making such a hedge is the traditional billhook. More often found in the hands of the butcher than the gardener, the billhook nevertheless has its horticultural uses.

The billhook was not only the hedge layer's favorite tool, it also had different applications around the garden.

The Billhook Family

As a basic cutting and chopping tool, the billhook is close kin to the Malay *panga*, the Cuban *machete* and the German forester's *Heppe*, as it is to the Italian *roncole*, the Nepalese *khukuri* or the North American fascine brush knife. All are multipurpose cutting tools and each has a hook-shaped blade mounted with rivets, a tang-and-ferrule joint, or a socket fitting and a short wooden handle.

In Britain the billhook developed many regional names: there was the socketed Suffolk, the Shropshire, the Ulverston, Banbury, Burton, Edinboro and Lincoln, each catering, as Roy Brigden put it, "for all requirements and prejudices" (*Agricultural Hand Tools*, Shire, 1983). Hedge-laying styles differ from region to region and different hedges require different billhooks. In stock-raising districts the farmer needs a robust structure that can bear the weight of a bullock leaning on it and spring back into shape afterwards. On the wind-whipped, sheep-cropped moor, however, the shepherd wants a barrier for his ewes, but not a target for the wind. In these districts a good hedge is said to be the one you could "throw a billhook through."

The living hedge, usually with a ditch on the neighbor's side, keeps browsing farm beasts out of the garden. Initially planted with thorn, oak, crab, hawthorn and holly saplings, the hedge is allowed to grow to a certain height before being cut and laid during the winter when the sap is

The wide variety of billhooks and "slashers" reflected the wide variety of regional hedge types.

down. To lay the hedge, stakes are cut (usually from the hedge saplings) and set, gently angled backwards on the hedge line (every hedge "travels," the angle of the stakes denoting the direction). The other saplings are then bent over, half cut through at the base so that they yield but will still grow, and "laid" around the stakes. Depending on the regional style, the finished hedge may be topped with a "whip" of hazel and left for 20 years at least before it needs relaying.

The Roman emperor-to-be Julius Caesar encountered such boundary hedges in France when he pushed through northern Europe during the Gallic wars around 50 BC. "These hedges present a barrier like a wall," he noted with some surprise. A thousand years later the French Normans were exporting their motte and bailey fortifications together with their defensive fences and ditches to Britain.

Living hedges form a natural network for wildlife and have additional advantages, as Thomas Tusser pointed out in 1573:

Euerie [every] hedge
Hath plenty of fewell and fruit.

Those who helped themselves to the kindling wood in someone else's hedge, however, could expect to be whipped until they "bled well" and be obliged to compensate the landowner for the damage to his property.

Following the English Acts of Enclosure, a deft seizure of land from the people, up to

TOOLS IN ACTION

Billhook Handling

Using a billhook requires skill and care if the user is not to end up in the emergency room. The blade must be sharp (hold it in a vice and run a whetstone along both edges) and the handle must fit comfortably in the hand. Handles are sometimes hooked to curve round the base of the palm. Whether cutting bean poles (stripping the side branches with the hook is known as *snedding*) or splitting hazel rods for a hurdle (a movable fence panel), the billhook is used in a chopping action. Wear gloves when handling hedge trees such as blackthorn, but always remove them when handling the billhook: it can slip from a gloved hand. Avoid leaving the hook stuck in a stump while working: instead, lay it flat on the ground.

200,000 miles (320,000 km) of new hedges were laid during the century leading up to 1850. For a century afterwards the laid hedge was as popular in the town villa garden as it was in the country.

There are alternatives to the hedge, many of which still rely on the use of the billhook. Gertrude Jekyll in the early 1900s championed the "true fence of the country and a thing of actual beauty": the oak post-and-rail fence. The oak posts and rails are cleft with a billhook and stitched together without nails. Similar traditions are followed elsewhere in Europe: one style popular in parts of Estonia involves weaving a thicket of hazel rods, or "ethers," between three horizontal poles. In some parts of Sweden and Russia, pine poles are slanted down at 45 degrees and bound to a timber frame with twists of bark; elsewhere, portable hurdles are woven out of timbers cleft from the local woods.

Then there is the picket or pale fence, from the Latin *palus*, "stake." Made from sawn boards rather than wood split with a billhook, the picket fence has come to typify suburban gardens from North America to North Island, New Zealand, although it originated with the "pale" of the Middle Ages. This was the nobleman's park, designed to keep the fallow deer in their proper place to be hunted. The medieval pale was made with the billhook from split or cleft oak stakes set in the ground and pegged to horizontal rails.

To prevent the deer from leaping the fence, a bank and ditch were added and the pale fence was run along the top of the bank.

All these fencing solutions made good use of the billhook and of local woodlands, a sure way of ensuring their continued survival. Miss Jekyll was left musing why anyone should bother to fence their garden with that popular Victorian boundary, the iron railing, when hedges, hurdles or her favorite "stout old oak post and rail", provided all the shelter from the weather that anyone could wish for.

The Dutch miniaturist Paulus Moreelse added his discreet signature to his 1630 painting, Vertumnus and Pomona, *to the hook-shaped knife Pomona carried.*

Rake

The multipurpose rake has a host of applications, from preparing the ground in the vegetable garden to raking up the autumn leaves. Although gardeners are wary of the misplaced rake smacking them on the head, in the Japanese Zen garden the *kumade* was believed to bring good luck.

Definition

A tool employed, like the farmer's harrow, to comb the ground for a variety of purposes.

Origin

The Roman gardener's rake was the *rastrum quadridens*, literally "a hoe with four prongs."

Carefully raked sand at the Japanese Kennin-ji Temple in Higashiyama, Kyoto.

In the early 1960s, an artist better known for his designs for interior furnishings was creating some curious gardens in his adopted city of New York. Isamu Noguchi, a benign-looking 60-year-old Japanese-American, had made a name for himself with his plans for domestic lamps and furniture. But he had a passion for ultramodern, dry landscaped gardens marked out with boulders and sand.

Designs such as his UNESCO garden or his subterranean garden outside New York's Chase Manhattan Bank, which opened in 1964, bridged East and West, inspired as they were by Japanese gardens. They harked back to traditional raked-sand and rock gardens of Japan's Zen Buddhists as well as that most basic gardening tool: the rake.

DEVOTIONAL RAKING

The carefully contrived Japanese gardens were made to mirror the natural landscape. They started to appear around 1,200 years ago, inspired by similar gardens in China, reflecting the Shinto belief that ancient stones and trees should be venerated as places imbued with spirits of the past, or *kami*. With the arrival of Zen Buddhism in the 12th century, such gardens offered appropriate places of tranquillity, a respite from what one commentator called "world's dust," and they evolved into contemplative sand-and-rock gardens.

The Zen garden was often placed close to a temple where devout monks, equipped with a special *sanon yo kumade* or sand rake, would sculpt the sand into restful patterns that swirled around the carefully placed mossy boulders. These strange, plant-free gardens became imbued with such profound symbolism that it was difficult to explain precisely what was being symbolized. In a sense, if you needed to ask, you had yet to reach the right stage of knowing. It was left to poets such as Edo's (Tokyo's) Matsuo Bashō, born in 1644, to offer the enigmatic haiku:

Calm and serene
The sound of a cicada
Penetrates the rock.

RAKE MAKING

There was a time when every country commune boasted its own rake maker. Together with the smith, the wheelwright, the potter and the basketmaker, the rake maker was an industrious craftsperson. Carpenters used the best regional woods for their rakes: seasoned ash, hickory, sycamore, elm, willow or, in the Far East, bamboo.

The traditional method involves cutting a row of wood tines and driving them into a rake head, which is secured to a shaft, also called a *stail* or *haft*. The rake maker kept a short metal tube, its edges razor-sharp, nailed over a hole on his shaving horse to fashion each dowel-shaped tine: the tines were cut by driving split timbers through the tube with a mallet.

The wire rake, useful for gathering together autumn leaves for composting, was also a handy scarifying tool to improve the aeration of the lawn.

Scarifying a Lawn

To scarify is to damage something by scratching or cutting it. Scarifying the lawn, however, is wholly beneficial. The close-cut turf, mown to within an inch of its life, looks neat from a distance, but the thick thatch of grass stems growing laterally (to escape the mower) clogs up this little ecosystem, restricting light and air and encouraging mosses and weeds to settle. The gardener can buy or rent a mechanical dethatcher. Fitted with small blades, it scarifies the grass, opening the turf to light and improving drainage. Alternatively, regular wire-raking will pull out dead matter (ideal for composting) and, on all but the toughest grasses, improve the condition of the lawn. An additional aeration in spring will also help: working systematically down the lawn, drive a garden fork into the turf.

For the wide-headed grass rake, the shaft was split part-way down and the base of the split bound and held with twine or strips of some pliable wood such as willow. These long-headed rakes with up to 30 tines were reserved for lawn work, clearing autumn leaves or scarifying the grass in spring. Short-headed rakes, reserved for the vegetable garden and used to reduce the soil to a fine tilth or to draw a seed drill, could be mounted on a single, broom-like shaft.

Wooden rakes gave way to iron and steel and eventually to wire and plastic while the shape of the head varied from a straight T, sometimes with slightly clawed tines, to a wide fan or crescent. One ingenious designer produced a wire rake that could be collapsed, like a fan, for better storage; another developed a rake where the pitch of the tines could be adjusted; and one old farmhand explained that when it came to running the hay-filled cart home from the meadow, the sides were always combed down with a long-handled rake to prevent the carter scattering loose hay on his way home.

For the Japanese sand rake, the *sanon yo kumade*, a one-piece head might be cut into a simple sawtooth shape and mounted on a cleft shaft. These are large and heavy, their very weight helping to shape the troughs and peaks of the sand waves. Alternatively, the head may be formed from a bamboo tube with small bamboo tines set in holes drilled into the head. The *sanon yo kumade* is designed not only to create the sand furrows, but when turned over, to soothe and smooth them away. So important is the rake that it has come to symbolize good luck and fortune, and during the traditional harvest festival, the *tori-no-ichi*, miniature rakes are decorated with streamers and hung

Raking is useful in smoothing the soil after digging, and in collecting weeds, stones, &c., and dragging them to one side, where they may easily be removed. An iron-toothed rake is generally used for the ground and a wooden one for collecting grass after mowing.

Jane Loudon, *Gardening for Ladies* (1840)

with sweets and coins. The lucky rakes are sold to passers-by accompanied by hand clapping by both stallholder and purchaser and chants of *Kanai anzen, shobai hanjo*—"Safe family, good fortune."

The lucky rake is an improvement on the rake customarily carried by Pesta, the ghostly Norwegian hag who was said to bring the plague to Scandinavian villages. When she brought her broom, all were condemned to be swept to their death; when she carried her rake, some at least could expect to be spared.

Japanese sand gardens had a significant influence on the West, not only in Isamu Noguchi's time but also in the early 20th century. A fashion for Japonaiserie was more pervasive than the cult of Chinoiserie, inspiring popular operas such as Gilbert and Sullivan's *The Mikado* and leading to the establishment of famous Japanese gardens such as Tully near Kildare in Ireland, Clingendael Park at The Hague in the Netherlands and the Nitobe Garden in British Columbia.

Mechanical Tiller

From the clod crusher to the
Bodenfräsen or earth grinder, the
mechanical tiller has undergone
a slow evolution in the garden.
The breakthrough came with the
New South Wales farmer
Arthur Howard and his
lightweight Rotovator.

Definition

A motorized implement with
rotating tines or blades to turn
and loosen soil.

Origin

From the Latin *mechanicus*,
"engineer," and Old English *tilian*,
"to treat or cure."

The notion of breaking down the soil to produce a finer tilth, and of scraping at, or chopping the earth to control unwanted plant growth, is as old as the first, ancient forms of the mattock and hoe. For generations, single-person, pushable, toothed cultivators, known also as wheel hoes, or push plows, were, and still are, available for smaller areas of planting. Larger areas with crops needed larger implements.

By 1800 the practicalities of steam power had consolidated Britain's basic industries such as mining, textiles and farming. The steam threshing machine had arrived in the farmyard in the 1830s, taking away the jobs of thousands of farm laborers. The job losses triggered rioting and, of the 2,000 country men subsequently brought to trial, nearly 500 were put on the convict ships bound for the Australian penal colonies. Another 19 were hanged.

William Crosskill's steam powered cultivator takes to the fields in the 1850s. Gardeners would wait another 60 years for its horticultural equivalent.

AGRICULTURAL ARMS RACE

But steam power was unstoppable and in 1839 it was already making its way into the fields. An engineer named William Howden demonstrated a portable steam engine capable of driving a variety of implements, to an astonished group of Lincolnshire farmers. Howden, however, produced only twelve of his "steamers" and two years later, in 1841, Ransomes of Ipswich took to the Liverpool Royal Show a portable steam engine capable of delivering 5 horsepower. (The Scots engineer James Watt had coined the term

"horsepower" to equate steam power to that of the draft horse.) A year later Ransomes had produced a self-propelled version, opening the way for steam-powered plowing and traction engines. They were to transform the world of farming from the 1870s onwards, and at the close of the 19th century, over 33,000 engines were at work across the landscape of Britain.

The year 1841 had also seen developments in rotary-action cultivation aimed at breaking up the soil. William Crosskill, a talented entrepreneur from Yorkshire with a penchant for manufacturing objects in iron, invented the fearsome-sounding Self-Cleaning Clod-Crusher and Roller, a massive, tined disc harrow that earned him a Gold Medal from the Royal Agricultural Society and the machine a place in the Great Exhibition of 1851.

Such agricultural developments were about to filter down to the more domestic needs of the individual and community gardeners: they would arrive in the form of the mechanized rotary tiller. The earliest functioning mechanical tillers, called *Bodenfräsen* or earth grinders, were designed and constructed by Konrad Victor von Meyenburg, a German engineer. The size of tractors, these formidable machines were self-propelled and used choppers that swung to and fro in the earth. Von Meyenburg developed spring-mounted tines that were flexible and enabled the machine to maneuver around obstructions such as stones, patenting the system in 1909. From 1912 to 1914 several prototypes were constructed and exhibited in Europe and America. Based on his designs, small rotary tillers were produced by other companies in Germany and Switzerland, but they were never robust enough to withstand the stony

soils of the United States and the tines were constantly breaking. It was left to a New South Wales farmer, Arthur Clifford Howard, to bridge the gap.

Howard's prototype rotary tiller started out attached to the engine of a steam-driven tractor on his farm at Gilgandra near Crookwell in 1912. He had been endeavoring to find an easier way of working the soil that did not expend so much human energy. Howard, a quiet yet determined individual, began experimenting with turning the earth by mechanical means while avoiding the kind of soil compression caused by the conventional plow. Using a selection of ransacked farm-machine components such as cogs and sprockets, and with his father's tractor as the power source, Howard fixed the blades of several hoes to the shaft of a disc cultivator. His first efforts sprayed the tilled earth out to one side because the discs spun too quickly. He adjusted the design with a set of L-shaped blades on a smaller rotor.

Work on the Howard mechanical tiller was brought to a halt with the advent of World War I. Unfit for military service owing to a motorcycle accident, Arthur Clifford Howard traveled to Britain to work on munitions and demonstrate his idea for "rotating hoe blades" to UK engineers. He failed to find a backer.

A relaxed New South Wales farmer with his new Howard cultivator. It promised to transform the back-breaking business of digging over new ground.

Using and Choosing

Before choosing a tiller, carefully consider the size of your garden and how much of it will need to be prepared. A newly made garden, or one with unbroken ground, calls for a rear-tine tiller with greater horsepower. Operating the machine with proper care requires preparatory clearing of the ground. Remove large stones and roots, tree stumps, rubbish and other potential obstructions that could make the tiller bounce out of control. Set the depth of the tines to a medium-to-deep adjustment. Use the lowest gear when breaking new ground to avoid burning out an over-worked engine.

Back in Australia in 1920 Howard completed the final tests and patented his rotary-hoe cultivator, now powered by an internal combustion engine mounted on the main frame and equipped with five hoes on a secondary frame.

Howard's trademark Rotavator became the mechanical workhorse of the 1950s residential gardening boom. Technical improvements included a gearbox for adjusting the forward speed while the tines continued to rotate at a constant speed, and a reverse gear.

C. W. Kelsey, an American automobile producer, had established an office in New York in 1932 and, registering the name Rototiller, he started to import Swiss, German and Danish machines. By 1945 Rototiller Inc. was producing smaller walk-behind cultivators instead of their large tractor-mounted machines. Two years later, Clayton Merry designed the Merry Tiller in Washington, a tiller that has since acquired legendary status, acknowledged for its durability and strong construction.

The first of these machines in Britain were manufactured by the Birmingham-based Wolseley Sheep Shearing Machine Company towards the end of the 1950s and found their place tilling the earth of community gardens throughout the country.

Many of the modern mechanical tillers are, in reality, tractors masquerading under another name, complete with a power take-off and a seat. There is a narrow dividing line between the power tillers and the walking tractors that have become the contemporary equivalent of the postwar cultivator and are used extensively in many developing countries. As one example, among the many inexpensive models on offer, the Chongqing Guangyun Agriculture Machinery Company in mainland China was marketing a 7-horsepower gasoline tiller with belt transmission in 2013. Their brochure claimed it had "steady quality."

Composter

Garden designers have come up
with all kinds of clever composters,
from bottomless barrels to plastic
bins, in a bid to transform garden
waste into good humus. And humus,
claim its advocates, is the best way
to green up any garden.

Definition

A wooden or plastic structure
designed to convert vegetable
matter into compost or humus.

Origin

Compost, from Latin *compositum*,
"a mixture," is a traditional way
to enrich the soil.

here were strange goings-on in the 1960s at the little quasi-religious community on the northeast coast of Scotland. The community's founders, a former Royal Air Force officer named Peter Caddy, his wife, Eileen, and their companion, Dorothy Maclean, were managing a productive vegetable garden despite the dubious nature of the local sandy soil. They attributed their success to the divine intervention of angelic beings, or *devas*, although more sceptical observers put it down to their wildly successful composting systems, which enriched the land.

The community, Findhorn, went on to become a major New Age foundation, while the craft of composting continues to attract an army of advocates from Seattle Public Utilities ("Composting is easy and a great way to recycle yard waste and kitchen scraps into a fertile sweet-smelling soil builder") to Auckland's straightforward GoodShit Compost Company ("Call us and see why, when composting, only GoodShit will do!").

Whether it is made on an industrial or a backyard scale, in plastic Dalek-like barrels or specially designed tumblers, *bokashi* buckets or New Zealanders (lidded boxes that operate on the conveyor-belt principle), or in a moldering

TOOLS IN ACTION

The Foolproof Compost Heap

Create a pair of rectangular bays in an open E-shape, preferably using recycled lumber. Close the front of each bay with boards lightly nailed in place so that they can be removed for easy access: a cross-support across the top of the bays prevents the side walls from splaying out. Spread a layer of old branches or a line of bricks at the base to aid air circulation within the heap. Fill one bay at a time, layering the vegetable waste, grass clippings and weeds between layers of garden soil, old compost or manure (manure will speed up the process). Cover with an insulation layer such as old carpet. When the compost has turned dark brown and crumbly it is ready for the garden. Avoid using meat or dairy waste (it will attract rodents).

heap at the end of the garden, the principles are the same: any pile of vegetable matter will be turned into humus. And humus is an essential organic matter that enriches the soil.

Provided we do not drive our soils too hard, the land will go on feeding us through the sunlit centuries when motoring is but a memory.

Lawrence D. Hills, *Organic Gardening* (1977)

TREASURED RUBBISH

The gardening writer James Shirley Hibberd stressed compost's virtues back in the 1860s: "You have one source of the very best manure in the household, and you must treasure every scrap of stinking rubbish, solid and liquid, and not waste so much as a dead cabbage leaf," he declared. Hibberd would have been horrified by the quantity of kitchen waste squandered during the latter half of the 20th century. No other civilization in history had consumed natural resources on such a scale, nor junked their waste in such a casual manner. Even as scientists warned that the gas was running low, backyard gardeners were throwing away kitchen waste while fertilizing their beans and begonias with store-bought, petroleum-based chemicals. According to the Mother Nature Network, the average American household of four was, in 2011, still dumping almost $600 worth of food every year. Yet the case for composting had been forcefully made over half a century earlier. Annie Francé-Harrar and Lady Eve Balfour, both born towards the end of the 19th century, were among its chief proponents. Francé-Harrar was an Austrian who began work on composting in Budapest after fleeing the war in Germany. She went on to work with the Mexican government, tackling soil erosion through composting, and in 1958 published her book *Humus: Soil Life and Fertility*. Eve Balfour, meanwhile, published *The Living Soil* in 1943, having started an organic farm at Haughley Green in Suffolk, England, in the 1920s. Three years later she co-founded the UK's Soil Association. Balfour acknowledged the work of an earlier contributor to the composting debate called Rudolf Steiner. Born in 1861, Steiner was devoting himself to the "science of the spirit," which he called anthroposophy. His teachings led to the founding of community villages devoted to the care of people with learning disabilities, an international network of schools based on experiential learning, and the bio-dynamic method of organic farming and gardening.

John Soper, a former agriculture adviser in Tanzania and the author of *Bio-Dynamic Gardening* (1983), summarized their thinking: "The earth breathes, it has a respiratory system, it has a pulse, it is sensitive and it has a skin." This earthly life form, he said, expanded and contracted like any breathing being, and he advocated that the gardener work in harmony with its rhythms, harvesting leaf vegetables and transplanting seedlings in the expanding mornings, for

Gardeners waste good compost when they throw their vegetable peelings away.

"Treasure every scrap of stinking rubbish," declared an early proponent of the craft of composting, the Victorian garden writer James Shirley Hibberd.

example, and sowing seeds, harvesting root crops and transplanting small plants during the evenings as the earth gently contracted.

The bio-dynamic movement continued to gather support, though even Steiner's supporters sometimes found his ideas difficult to follow.

The same could not be said of another composting hero, Albert Howard. Howard had warned in the introduction to *An Agricultural Testament* (1940): "The agriculture of ancient Rome failed because it was unable to maintain the soil in a fertile condition. The farmers of the West are repeating the mistakes made by

Imperial Rome. How long will the supremacy of the West endure?"

Howard's work, which inspired others including Jerome Rodale, the New York author who propagated the organic message in America, was based on his experiences in India. He had departed for the British colony intending to educate the natives on the benefits of Western agriculture; he returned home convinced by Indian farmers themselves that "Nature's Methods" for maintaining soil fertility, adding back the humus from composted vegetable and animal wastes, were the best. Composting was not a trifling matter for weekend gardeners, and mankind needed to regulate its affairs so that its chief possession, "the fertility of the soil," was preserved: indeed, the future of civilization depended on it.

Howard was a practical man and he provided the gardener with a practical tool to create humus: the Indore container composter. A slatted wooden structure was built and gradually filled with 6 in. (150 mm) layers of compostable materials sandwiched between 2 in. (50 mm) of manure and a sprinkling of soil until the heap reached 5 ft. (1.5 m). The pile was kept moist, turned at six weeks and again at twelve weeks, and was ready in three months (longer in winter).

The recipe could be speeded up by mixing the compost with manure and turning the heap only once. Similar methods involved laying the biodegradable material directly on the soil and digging or forking it in; or reducing the depth of all materials to around 2 in. (50 mm). One environmentally unfriendly but effective method involved running them over with the lawn mower before heaping them into a 5 ft. (1.5 m) high pile. Turned every three days, this intensive composting was said to mature in three weeks.

There were rival methods including the *bokashi* system ("fermentation" in Japanese) that pickled all the kitchen waste with the assistance of a bacterial bran in a special closed bucket. Then there was the apparent magic of comfrey.

In the 1950s the freelance journalist Lawrence D. Hills was in the vanguard of the organic food-growing movement. As he would write in 1977: "Some [gardeners] change [to organics] on ethical grounds to stop pollution harmful to birds, bees and men, others to save money, since it is easy even at today's vegetable prices to spend more on chemicals than you save when growing your own food."

Interested in experiments with comfrey conducted by 19th-century Quaker Henry Doubleday, Hills trialled the plant as an organic aid. He had rented land at Bocking near Braintree in Essex, England, and was soon writing in his classic *Organic Gardening*: "Comfrey is so rich in protein . . . that it is a kind of instant compost." His leading variety (he called it Bocking 14) was a hybrid between *Symphytum asperum* from Russia and the wild *S. officinale*, the herbalists' comfrey. Hills went on to found an organization (named after Henry Doubleday) to spread the word. Eventually it became Garden Organic, under the patronage of the heir to the British throne, Prince Charles.

Comfrey was judged one of the most important herbs for the organic gardeners because of its composting qualities.

Hotbed

"Be avaricious for manure," counseled one 19th-century garden writer in the days when reluctant domestics were sent out into the street with shovel and bucket to retrieve the leavings of passing dray horses. The author was writing about one of the gardener's most useful tools: the manure heap.

T here was a time when most farmsteads boasted a midden heap from which the village gardener sourced his manure. The industrialization of farming and the 19th-century drift of populations from country to town, however, left city gardeners trying ever more inventive ways to enliven the plot. There was seaweed and spoiled hay, spent tanners' bark, pilchards (a small river fish) and whale blubber, horn and bone, hair and rags, blood, urine and coral, soot, house sweepings and soap makers' waste–not to mention pigeon, fowl, cattle, oxen, sheep, deer, camel and horse dung. The finest horticultural horse dung, according to Spain's Moorish gardeners, came from corn-fed stallions. And there was no point wasting good urine: the 11th-century Ibn Bassal recommended that laborers be persuaded to pee on the midden heap. (The Roman writer Columella went further, recommending that urine be stored for up to six months before being mixed with old oil lees and watered around the olive trees.)

Guano mining was big business in the 19th century and provided the gardener with a powerful fertilizer.

PERUVIAN GUANO

"The most stimulating of all known manures" was Peruvian guano, according to James Shirley Hibberd, who advised that it be sparingly applied: "In using guano mistakes are frequently made." Guano, sometimes erroneously described as "monkey muck," was seabird manure extracted by the boatload from islands in the Pacific Ocean and anywhere else where large colonies of seabirds gathered. Guano mining was one of the boom businesses of the 19th century and surviving stocks occasionally became the focus of international disputes. When American sailors landed their barque on one of the guano-rich Lacepede Islands off the western coast of Australia in 1876 and raised the Stars and Stripes, it sparked an outcry. Western Australia was obliged to send a representative out to reassert its colonial rights over this mountain of seabird dung.

Aside from applications of guano, soil fertility was also enhanced by "night soil," that euphemism for the contents of the outhouse. In the Middle Ages, *gongfermors* (*gong* being a privy and *fermors* from Old English *feormian*, "to cleanse") were responsible for carting away the contents of the privy by night. Centuries later it was still a job for the night shift as

Hotbeds used by Gilbert White. A pit is filled with horse manure and topped with loamy soil. It ferments and produces a high temperature which aids seed germination.

one American, Carl Sandburg, explained in John Pudsey's *The Smallest Room* (1954): "About once a year, a Negro we called Mister Elsey would come with his wagon and clean the vault of our privy. His work was always done at night. He came and went like a shadow in the moon."

Night soil was dug in or spread out on farmland and gardens, since no one was in any doubt about the benefits of manure. (As far back as the mid-1600s, the Protestant and Leveler Gerard Winstanley noted: "If the wasteland of England were manured by her children it would become in a few years the richest, the strongest and the most flourishing country in the world.")

Gardeners were also learning to harness the heat of the midden heap. The anaerobic activity in a pile of fresh manure mixed with straw generated natural heat and gardeners discovered how to box the manure into a hotbed and grow an early crop of vegetables on the top. As the clergyman and naturalist Reverend Gilbert White wrote in his *Garden Kalendar* (1756): "Jan 23. Made an hot-bed on the dung hill in the yard with Mr. Johnson's frames, for white mustard and cress." Later he records: "Hot-bed works very well. Hard frost for two or three days: now ground covered with snow." Two days later, his hotbed was still working well: "On this day, which was very bright, the sun shone very warm on the Hot-bed from a quarter before nine to three quarters after two. Very hard frost."

The action of the hotbed, which had been benefiting North Americans since the first record of such a device in Virginia in 1773, would lead to the development of the heated glasshouse. One, installed on the Earl of Derby's estate in England in the 1790s, used the heat generated, not from manure, but from tanbark. Tree bark was used by tanners to condition their animal hides, and the waste bark or mulch could, when piled in a heap, produce a generous amount of heat. The Earl's gardener, who was trying to raise melons and pineapples, boosted the heat even further by running steam through perforated pipes in the bark pile. The technique led to the development of steam-heated pipes—usually involving a

An amateur gardener's hotbed constructed from a glass a timber frame over a bed of earth and manure.

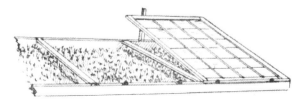

1 in. (25 mm) steam pipe passing through an 8 in. (200 mm) water-filled pipe—in the 19th-century glasshouse.

Manure was also used in liquid form. One professional gardener, who worked on Welsh country estates in the 1930s, described how he would take an old tar barrel, burn out the remaining tar and fill it with water. "We used to scrounge round all the farms for when the farmers were *dagging*, that was cleaning the rear ends of lambs and sheep before shearing. We used to collect all those, put 'em in a hessian sack and hang them in the barrel. That was our liquid feed. You couldn't beat it."

Science, however, was less certain about why it benefited the soil. In his search for an answer, Cardinal Nicolas of Cusa, a Renaissance mathematician who died in Italy in 1464, tried weighing samples of soil before and after they had raised a crop. Finding little difference in the weights, he rightly concluded that water played a significant part in the process. He also decided that plants absorbed some nourishment from the air.

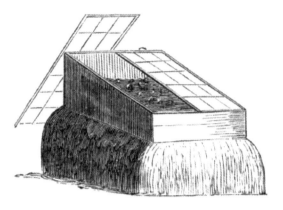

A mound of manure can be used to harness a heap of heat for tender plants.

Soil, it seemed, was simply a growing medium. When, after producing a crop of plants, fertility fell, it required only a nourishing dose of manure to restore it to active service. Horticulturists, however, were advised never to rest on their laurels: "Always keep your mind in firm conviction that your ground is in an impoverished state," recommended a stern Hibberd. "When well decayed they [manures] ameliorate the soil, improve its texture, and enable a crop to withstand drought for a longer time than in soils that have not been manured." Hibberd,

Ammonia, potash, soda, salt, and other chemical principles are essential to the fertilization of the soil. It is only necessary to remember, that whatever will putrefy and become obnoxious if exposed to the air, loses all its obnoxious qualities the moment it is mixed with the soil, which is a natural deodorizer, and all such substances are powerful in stimulating and feeding vegetable growth.

James Shirley Hibberd, *Profitable Gardening* (1863)

TOOLS IN ACTION

Recreating a 19th-Century Hotbed

The mist rising from a compost heap shows how much heat can be generated by the anaerobic activity inside. Gardeners learned to use these higher temperatures to raise early crops such as salad greens by growing them in soil in a frame placed over a bed of manure. Fresh straw manure is first forked into the base of the bed, tamped down to 6–7.5 in. (150–190 mm) deep and overlaid with around 1 in. (25 mm) of a soil growing medium garden compost, or a mixture of the two. A frame or box bed, built of wood and lined with some form of insulation, such as cardboard, is placed over the bed. Allow the heap to heat up for about a week before sowing seeds or planting seedlings. Use a thermometer to check that the bed does not overheat: cool it by watering if the temperature rises above 24°C (75°F). The hotbed can be set up in a glasshouse or outdoors: outdoor boxes should be fitted with a loose lid, closed during cold nights.

who was writing in the 1860s, was also frustrated by those traditionalists who were set against the addition of "ammonia, potash, soda, salt and other chemical principles . . . essential to the fertilization of the soil."

This was an oblique reference to the work of the man who was about to come up with the idea of meat extracts (which led to the invention of the bouillon cube): Justus von Liebig. A German scientist, Liebig had discovered that plants took in carbonic acid, water, ammonia, potassium, calcium, magnesium, phosphate and sulphate, converting them into starch, sugar, fat and proteins. And he realized that animals, which ate and excreted plants, returned these base elements to the soil.

Liebig's mission to uncover the secrets of the soil was attributed to the teenage trauma of having lived through the hungry year of 1816. (Liebig was born in 1803.) During that year the death rate soared and there were desperate food shortages across the northern hemisphere after a volcanic eruption in Indonesia, combined with a record low in solar activity. The resulting climatic conditions saw snow fall in New York during June and glacial flooding in the Swiss cantons. Farmers and gardeners struggled to grow anything during what turned out to be the last major food crisis experienced by the Western world. Liebig's research led to the development of artificial fertilizer, which has, so far, prevented a crisis similar to that of 1816.

Latin

The use of a dead Mediterranean language, Latin, to describe the gardener's plants has always attracted critics. Nevertheless, Carl Linnaeus's binomial system for naming plants and flowers has given botanists and gardeners an international *lingua franca* for more than 200 years.

Definition

Latin nomenclature is used to give an accurate and internationally accepted description of all living things.

Origin

Latin, the international language of its day, was adopted by Carl Linnaeus in the 18th century to describe the plant world.

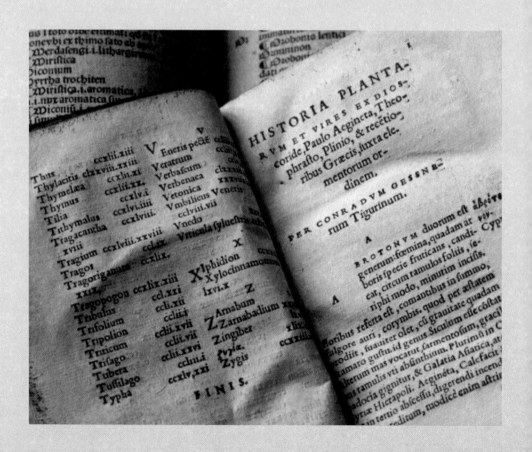

T he worlds of botany and horticulture come together in the language of plants. Latin might not fall into the more obvious category of physical tools, but nevertheless, it provides the gardener with an accurate and unique description of every plant and a language that can be shared and understood by gardeners from Kazakhstan to Kentucky.

Latin did not please everyone: "Why," wails one garden blogger, "must I describe plants in an unpronounceable language that died out almost 2,000 years ago? An Auckland farmer doesn't order a *Bos hornabreviensis*: he orders a Shorthorn cow."

The long answer is that we have saved a forest's worth of paper by adopting the basic, two-word system invented by a Swedish botanist who had a curious preoccupation with plant sexuality. The short answer is: it works.

Consider the historic ship that sailed to America with 120 pilgrims on board in 1620, the *Mayflower*. The ship set off trailing floral confusion in its wake. The vessel took its name from the delightful *souci de l'eau*, as it was known in France, or the marsh marigold, pollyblob, May-blob and kingcup as it was called in different parts of Britain. In fact, the mayflower boasts almost 300 different vernacular names, yet the "mayflower" of North America (which is also the state plant of Massachusetts) is a different thing altogether. This is the trailing arbutus, which earned its common name from the Pilgrim Fathers who first trampled across

Horticultural Latin cleared up the confusion over the original mayflower, Caltha palustris, *illustrated here by Otto Wilhelm Thomé in his 1885* Flora von Deutschland.

wild beds of it when they landed in New England in 1620. In the soothing lexicon of Latin, all confusion comes to an end: the many-named European mayflower earned its official title *Caltha palustris* L. in 1753. The Massachusetts mayflower was named *Epigaea repens* L.

The "L." refers to the man who set the standard calibration for the garden thermometer (*see* p. 164), learned to grow bananas in the Netherlands, set the standards for modern

Loving Latin

Latin is more precise and practical than poetic in its use of plant names. The descriptions *columnaris*, for example, or *sempervirens*, tell us that the plant is tall or evergreen, respectively. And it is useful to understand the different elements of the plant names together with the conventions when recording them. The Family names end in *–aceae* and are italicized: *Fabaceae*. Then comes the genus (plural: genera). Being a Latin noun, it has a gender—masculine, feminine or neuter: *Pisum*, for example, is neuter. Next is the species, written in lower case and usually in the form of an adjective agreeing with the noun: *Pisum sativum* (*sativum* is the neuter form of *sativus*, "cultivated"). A distinct variant of the species is the subspecies (subsp.), while the term "variety" (abbreviated to "var.") recognizes slight variations: *Pisum sativum* var. *macrocarpon*, for example, refers to the snow pea.

botanical gardens such as the Eden Project, and devised a system of classifying every living form—Carl Linnaeus. Linnaeus was a modest man. He stipulated that after his death his family should "entertain nobody at my funeral, and accept no condolences"; yet, in January 1778, even the King of Sweden came to pay his respects at Uppsala Cathedral where Linnaeus was interred.

Linnaeus was born in 1707, the oldest child of Nils Ingemarsson Linnaeus. A parish priest, Nils would walk with his five-year-old son through the countryside instructing him in the correct names of local plants. Later he persuaded Carl to study medicine at his old university, Uppsala. It was here that Carl befriended fellow

The Swedish botanist Carl Linnaeus made himself responsible for classifying all living things after his friend and colleague drowned in 1735.

student Peter Artedi, and the two young men
hatched an ambitious plan to classify all of
God's plants and creatures in a systematic
fashion. Promising each other that whoever
finished first would come to the aid of his
friend, Linnaeus and Artedi began working
through the animal and plant kingdoms. When
Artedi fell into an Amsterdam canal and
drowned in 1735, Linnaeus went on alone.

Around the time of Christ, the Greek
physician Dioscorides had diligently named
some 500 plants in his *De Materia Medica*
(*On Medicine*), but it had taken almost 1,000
years for his work to be disseminated, first
through the Arabic, and then the Latin, worlds.
Latin was then the language of learning and
was applied to plants and flowers. The Greek
polymath Theophrastus, for example, identified
the pretty scented pink as *Dianthus* ("God's
flower"), but to distinguish it from over 300
other species of pink, the clove pink became
*Dianthus floribus solitaris, squamis calycinis
subovatis brevissimis, corollis crenatis* ("the
dianthus with solitary flowers and short
inverted egg-shaped scaled calyces together
with crown-shaped corollas"). Linnaeus
chopped it short. It was to be the clove-flowered
pink, *Dianthus carophyllus*. It remains so today.

To begin with, Linnaeus was confronted
with a variety of often conflicting Latin titles,
including those of Dioscorides, and vernacular
names. He and the other botanists of the day

Dioscorides' Materia Medica *listed many plants with
healing properties, including those that could treat a dog
bite, illustrated in this 13th century Arabic translation.*

agreed that flowers belonged to the plant
kingdom. Linnaeus also understood that
different plants fell into natural orders (later
referred to as "families"): onions, leeks and garlic,
for example, to the *Amaryllidaceae* family, or
cabbages and cauliflowers to the *Brassicaceae*
family. Within the natural orders were genera
and within the genera were species.

Linnaeus took the 5,900 plant species then
known and gave each a two-word name. The
names were published in his *Species Plantarum*
(*The Species of Plants)* in 1753 and provided an
international starting point for all the old

botanical names. Having subdivided the plant families into their distinct groups, or genera, he employed the genus for the first name (*Pastinaca* from the Latin *pascere* ("to feed") for the parsnip, for example, or *Pisum* for the pea) and the species for the second name. So the garden pea and the sweet pea, which shared the distinctive-shaped flower or standard and that therefore belonged to the same family (*Leguminosae* or *Fabaceae*), became *Pisum sativum* and *Lathyrus odoratus*, both distinct from the tawny pea, *P. fulvum*. Finally, each species could be further divided to include cultivated varieties or cultivars such as *Erica carnea* "Myretoun Ruby," where the name is placed in inverted commas.

Condemned for its "licentious method" of classification, Linnaeus's system has stood the test of time.

LICENTIOUS LINNAEUS

Linnaeus classified the genus and species of plants according to the number of stamens and stigmas, a sexual system of classification. (One of his proposed species, for example, was *Polygamia necessaria*, or as he put it: "when the married females are barren and the concubines fertile.") He also frequently employed the names of friends, like the St. Petersburg academic Johann Georg Siegesbeck, after whom he named the genus *Siegesbeckia*. Siegesbeck, however, rounded on his friend, denouncing Linnaeus's work as "lewd." How, demanded Siegesbeck, could young people be taught "so licentious a method" of classification? Despite what Siegesbeck thought of Linnaeus's system, it was adopted universally. Linnaeus became a household name, and the bluebell or *jacinthe des prés*, wild hyacinth, bell bottle or fairy flower was nailed down as *Hyacinthus non-scripta*. It subsequently became *Scilla non-scripta* and then *Endymion non-scripta* before settling down as *Hyacinthoides non-scripta*—for the business of naming plants continued long after Linnaeus, currently through the International Code of Nomenclature for Cultivated Plants. The continued use of Latin arose out of an international horticultural congress in Brussels in 1864 when the botanist Alphonse de Candolle proposed that the language should apply to all wild species and varieties. The congress did make one concession to modernity: cultivated varieties that came out of the garden could be given a fancy or non-Latin name.

Raised Bed

Adopted by Aztecs and early
American colonists alike, the elevated
bed has served the gardener's needs
for centuries. While edging materials
have ranged from lead to animal
bones, the use of inverted beer
bottles, according to one Victorian
authority, was simply in bad taste.

Definition
A growing plot raised and enclosed by
a retaining border of timber or stone to
prevent the earth from spilling out.

Origin
"Raise" from Old Norse *reisa* "to raise";
"bed" from Old English *bedd*, meaning
a sleeping or growing place for plants.

I ndigenous peoples across the world have arrived at their own particular localized forms of raised-bed gardening, dictated by geographical location, weather patterns and lifestyle. From A.D. 1000 to 1200, long before the Europeans arrived, the Mill Creek Indians of northwest Ohio lived on both hunting and agriculture. With bone tools fashioned from the shoulder blades of bison or elk they tended their corn crops in gardens of unbordered, mounded earth, raised just high enough to prevent flood and frost damage, but also to extend the growing season—a method that came to be called the ridged-field system.

It was a practice that would be taken up by the Shakers, the followers of the independently-minded and remarkable Ann Lee, who in the 1770s had taken her flock from England to New England, Kentucky, Ohio and Indiana. The *Gardener's Manual* for 1835, written by the Shaker seed merchant Charles Crosman, highlighted the sect's passion for agriculture and advocated much the same approach for his readers. The author noted that "some sow or transplant on moderate ridges, and others on very high ridges," observing that the hollows and ditches between these raised mounds "were kept open to drain off the superabundant water in wet weather." (Strict adherents to the Shaker philosophy of simplicity in all things refused to raise flowers, because they regarded them as unnecessarily showy.)

The raised bed tradition continues in America to this day, although the particular method used came originally from the East. Peter Chan, a well-known gardener in the Oregon area, made his raised beds in the same way as those traditionally formed in China for over forty centuries: loose mounds with compacted, sloping edges, and no wooden surround whatsoever.

Elevating the earth for growth was a method that pre-dated cropping in rows. The Egyptians, the Greeks, the Romans and many older European cultures practiced various forms of raised-bed planting. In medieval Britain, rectangular beds were edged with basket-style, low woven hurdles and other forms of timber, tiles or, in some cases, an array of animal bones cleaned and aligned, with the knuckle joints standing uppermost as ornamentation.

The garden writer Charles Crossman extolled the virtues of growing plants on raised ridges, as practiced by Quakers, in his 6 cent Gardener's Manual *of 1835.*

CANOE GARDENERS

At the same time, in the Xochimilco lake basin in Mexico's central valley, a land reclamation system of raised-platform gardening was already in place. It began with the lowland Mayan peoples and later became subsumed within the extending Aztec empire. The system entailed land creation on top of lake wetlands.

The name given to this system was *chinampa*, from the Nahuatl meaning "reed basket" and "upon." It was an apt terminology: rectangles for planting were enclosed using juniper trees (*Juniperus* spp.) or fences woven from tule reed (*Schoenoplectus acutus*). Which were layered systematically with silt-rich mud from the bed of the lake, aquatic vegetation, silage and manure. Capillary action between these layers retained moisture and facilitated the cycle of nutrients to form a fertile soil that was alive with organisms and capable of sustaining up to seven crops a year.

Though they are often referred to as "floating gardens," the *chinampas* were fixed, stationary

The floating gardens or chinampas *of the Mayan people were destroyed within four years of the Spanish conquest of the Aztec nation in 1519.*

structures anchored to the lake bottom. Thousands of islands covering an area of 463 square miles (1200 km²) were created, connected by causeways and tended by gardeners who traveled about by canoe. In 1519 Hernán Cortés, the Spanish *conquistador* who would bring down the Aztec empire, witnessed all this at Tenochtitlán (present-day Mexico City). Within four years the *chinampa* system had been destroyed, the city laid waste and its inhabitants killed or in slavery.

TOOLS IN ACTION

Establishing Raised Beds

Despite the initial outlay, elevated beds offer the gardener significant advantages. Access to the planting is easier, at a comfortable level for working (a height of 27 in. (680 mm) is suitable for wheelchair users), drainage and soil structure are improved, and the plot is easier to control. In a raised bed you can grow a denser crop—as much as 10 times the amount, compared to the same area at ground level. Diseases and pests are moderated, the earth warms up earlier in the year, the growing season is extended, and the subsequent yields are higher. A 5 in. (130 mm) border of bare gravel or mulched earth around the bed will make mowing simpler.

Nevertheless, the longstanding application of raised-bed planting across Mesoamerica was a testament to the flexibility of the principle, which was adapted to differing climatic conditions and terrains. At 12,500 ft. (3,810 m) above sea level on Peru's border with Bolivia, for example, local people still use a growing technique established around 300 BC, pre-dating the Inca empire. Set out as a combination of long, narrow raised beds with irrigation channels between, the Quechua Indians called the system *waru waru*.

In 17th-century England, gardeners were advised to border their elevated beds with lead, "cut to the breadth of foure fingers, bowing the lower edge a little outward," or with "oaken inch boards four or five inches [100–130 mm] broad." This advice came from the apothecary John Parkinson, herbalist to Charles I, who described, with some revulsion, how Belgian and Dutch gardeners walled beds with ranks of jawbones.

The gardens of the early American colonists, were derived from the medieval and Tudor patterns of homeland Europe. From around 1600 to 1775 the early immigrants in Plymouth and rural New England planted or sowed onions, leeks, carrots, cabbage and radishes, growing them close to the house in rectangular or square raised beds, walled in by borders of cut saplings set in the ground. In order to maximize the space, the plants were grown in closely-spaced rows and all within the characteristic white picket fence (*see* p. 72), to protect them from browsing animals.

Modern raised beds, supported here with a framework of recycled railway sleepers.

Across the world, gardeners have demonstrated their customary ingenuity in creating raised beds. On the Canary Islands, volcanic rocks were formed into crescent shapes to protect tender plants; native plants in the 1,000-acre (400 ha) valley garden at Limahuli in Hawaii, set up in 1967, are grown in raised beds built from volcanic moss rock, a basalt that was also used to form walls by early European colonists in New Zealand.

Today's gardeners can turn to a range of reusable materials for their raised beds: reclaimed stone, blocks and bricks, straw bales, spare lumber, wooden posts, metal or plastic landscaping edging for curved structures, a child's paddling pool, old milk crates, recycled truck tires or galvanized water troughs.

Edward Kemp, one of Victorian England's leading garden designers with clients mainly among the *nouveau riche*, offered his own advice in the first edition of *How to Lay Out a Garden*, published in 1858: "As a rule, all sorts of freak edgings are to be eschewed, as, for example, the wire edgings in vogue fifty years ago, or edgings of whitewashed stones, or of bricks standing uncertainly on their corners." He recommended that the gardener avoid the use of "inverted beer bottles, or other convenient debris," which "no matter how curious and striking can hardly be said to be ornamental or in good taste."

Soil Sieve

The soil sieve or riddle is a simple tool that fell from favor with the introduction of industrially produced seeding and potting composts. However, its design has changed little over several thousand years and it still merits a place in any garden tool shed.

Definition

A container, usually circular in shape, with a bottom made of wire mesh.

Origin

"Sieve" is from Old English *sife*; "riddle," from the late 12th-century *hriddel*, meaning a coarse-meshed sieve.

D ue to the nature of its woven screen structure, the sieve was closely connected to the ancient craft of basketry (*see* p. 28). The principal use of a sieve was to sift soil to remove stones and other debris. As a basic device for separating elements one from another, it has been around for a very long time: one example discovered in a burial barrow in Saxony dates back thirty centuries.

The nature of any given tool is linked to the technology available to a particular culture. That early Saxon sieve was cast in bronze and bore a handle ornamented with the horns of a bull. In Britain, containers formed from steam-bent wood have been made since the Iron Age. This traditional approach to the craft continued to the present day. Sieve maker Mike Turnock was still making wooden horticultural riddles at his workshop in the Peak District in 2005. Producing some 120 sieves a week, he cut and steamed rims of beech wood, bending them into perfectly circular frames and, as his father had done before him, weaving the metal screens, using a crook-shaped tool to thread the wire. In his father's time, in the 1950s, the railways had been his principal customer, using large sieves to screen the ballast that was laid between the tracks.

The English author Jane Loudon, in another of her popular manuals, *The Ladies' Companion to the Flower Garden* (1865), stipulated that sieves were "necessary in gardening to separate the stones and coarser particles from the mold to be used for potting and also for cleaning seeds." She also used her sieve on gravel pathways to filter out any stones larger than gooseberries.

Jane Loudon was a fastidious gardener and she offered plenty of practical, no-nonsense advice: "The wires or *toile métallique*, through which the mold is to pass, should be firmly attached to the rim, the holes or interstices not being more than the fourth of an inch [6 mm] in diameter."

TOOLS IN ACTION

Seed Compost

A seed-starting mixture for general use can begin with three parts humus from your compost heap. Wearing gloves, work this through a fine-meshed sieve. Some growers recommend heating this screened soil to 65°C (150°F) in order to kill weed seeds and pathogens. One part coir, a fibrous material produced from coconut husks, is a sustainable substitute for peat that provides a good base for a seed-starting mix. Add one part vermiculite, made from the mineral mica and capable of absorbing and retaining moisture several times its own weight, and one part of the additive perlite, a form of inert, volcanic glass that facilitates good drainage.

DIVINATION BY SIEVE

The sieve is an ancient tool, once central to the strange world of coscinomancy. This divination practice, followed in New England in the 17th century but with origins that reached back to the ancient Greeks, relied on the sieve and the points of a pair of shears. According to the 15th-century German magician-alchemist Cornelius Agrippa, the movement of the riddle was controlled by a demon. This practice prompted the English playwright Ben Jonson to refer to "searching for things lost with a sieve and shears" in *The Alchemist* (1610).

The average pot boy employed on the English estate gardens of the 19th century had a more down-to-earth purpose for the sieve: he had to supply shards from broken-up terracotta flower-pots for use as drainage in the base of planting containers. He would break up the pots with a hammer , then run the pieces through the sieve.

Another of the pot boy's tasks was to prepare the fine seed-starting mixture, known as seed compost. For this purpose the pots could be smashed and sieved into a fine-grained powder. This was incorporated into the seed compost because it helped in retaining water, although Jane Loudon warned that "sifting however, should be used with caution; as most plants thrive better when the particles of soil are not too fine."

The term "compost" (from the Old French *composte*, or "mixture") has caused constant confusion in amateur horticultural circles since it refers both to the humus produced by the compost heap (*see* p. 84) and the seed or potting mixtures used to grow plants. Home-made recipes for seed and growing composts differ, but most seed compost is composed of finely sieved soil or loam, and sand (or crock powder), with the addition of sieved humus, such as leaf mould or peat for potting. Sometimes loam would be made from grass turfs, banked upside down for as long as it took them to rot.

The introduction of mechanically made seed and potting mixtures in the postwar years not only brought to an end the thrifty gardeners' practice of making their own, it also saw huge damage done to peatlands. Peat bogs are important wildlife wetlands that contain "locked-in" carbon which is released as a greenhouse gas when used in the garden. And yet by 2013, the gardener's addiction to peat-based compost showed no signs of abating, despite the loss of 94% of Britain's lowland peat bogs.

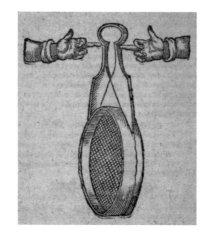

More than a soil sifter, the sieve could divine the future according to the ancient art of coscinomancy.

Radio

Some gardeners regard music as an
essential gardening tool. The 17th-
century writer John Evelyn believed
garden music to be an absolute
necessity, while the 20th-century
Dorothy Retallack was convinced that
plants themselves needed music—
so long as it was "improving."

ONE OF MARCONI'S EARLIEST EXPERIMENTS

The advent of the "wireless" radio was a boon to the gardener who enjoyed a bit of background music and a curse to those who preferred to garden in silence. A student at Temple Buell College in Denver, Colorado, Dorothy Retallack, however, was convinced that plants could hear.

She initially grew three groups of plants: the first in silence, the second subject to an intermittent tone, the third to the same tone, but played constantly for eight hours a day. The third group of plants died within two weeks. Convinced that her laboratory plants could hear, she repeated the experiment with two collections of plants, one subjected to a three-hour dose of music from a middle-of-the-road radio station, the second to the output of a rock station. Those exposed to rock music expired, while the middle-of-the-road listeners thrived and even grew yearningly towards the music source.

Dorothy continued her experiments and published her findings in *The Sound of Music and Plants* (1973). When Christopher Bird and Peter Tompkins highlighted her work in their *Secret Life of Plants* (also 1973), gardeners around the world were intrigued. It prompted scientists from the New York Botanical Garden to expose marigolds (*Tagetes erecta*) to a musical feast. They found no evidence of growth improvement.

One of the inventors of the wireless radio, Guglielmo Marconi, conducts his preliminary experiments with wireless telegraphy at the Villa Griffone, Pontecchio.

If garden plants were to benefit from the sound of music, they would need to be able to "hear," or at least detect musical sound waves. Plant scientist Daniel Chamovitz (*What a Plant Knows*, 2012), concluded that plants, although sensitive to touch and odor, were deaf to music. His conclusions matched those of Charles Darwin, the 19th-century naturalist and amateur bassoonist who, out of curiosity, played his bassoon to a mimosa (*Mimosa pudica*) to see if the sound caused the leaves to close as they did when touched. They did not, and Darwin concluded that his work was "a fool's experiment."

Mimosa pudica *(sensitive plant).*

THE HARMONIOUS GARDEN

The 17th-century English gardener John Evelyn, however, considered music in the garden "an absolute necessity"—but for his garden guests rather than the plants. Evelyn was a wealthy and well-traveled horticulturist and arboriculturist, and the author of *Sylva, or a Discourse of Forest Trees* (1664). Having married well, he inherited

TOOLS IN ACTION
Wireless Gardening

The radio brought horticultural wisdom to amateur gardeners long before the rise of television gardening programs. Yet even as radio shows such as the BBC's *Gardeners' Question Time* celebrated its 60th anniversary in 2007—a program born out of the wartime Dig For Victory campaign—the world of electronic communications had taken a new turn. A swipe of the screen or the touch of a button allowed gardeners to consult specialist websites, identify plants, download landscape design apps, source vintage garden tools or learn more about everything horticultural from permaculture to planters. While users should take precautions to protect themselves online, this brave new world of garden blogs and forums represents the most radical advance in communications since the advent of radio broadcasting.

the manor house and garden at Sayes Court in Deptford on the outskirts of London, and began his never-finished *Elysium Britannicum*, aimed not at the country "cabbage planter" or suburban "cockney-planters," but at the gentry, those "princes, noblemen and great persons" who could afford at least a 70-acre (30 ha) garden with knot gardens, parterres, groves, pavilions and labyrinths surrounded by a decent brick wall.

He wrote in his diary of a visit to the Prince's Court at The Hague in October 1641, where he admired the "Menage" and "a most sweet and delicious garden, where was another grot of small neat and costly materials full of noble statues, and entertaining us with artificial music."

Automated music makers were powered by wind or water. None quite excelled the great organ that had been unveiled at the Villa d'Este in Tivoli 60 years earlier, which was powered by a mighty fall of water. More subtle garden sounds were wind-powered chimes and bells. The Romans employed their garden *tintinnabula* or wind chimes to keep peace with the gods, in much the same way as Oriental gardeners employed garden wind bells to improve the *feng shui* (the practice of encouraging the most productive interaction between a person and their immediate environment) and maybe harvest a bit of good luck.

For all his rushing about Europe, Evelyn recognized that constant and elusive feature that gardening shared with music: the capacity to engender a sense of peace and quiet.

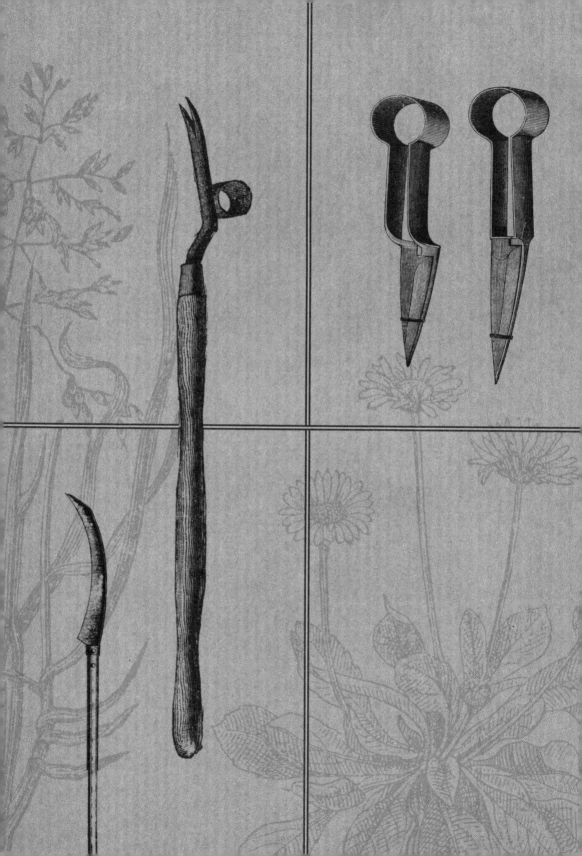

The
Lawn

T he casual lawn, or
what Pliny the Younger
described as the "little meadow
in the garden," has evolved
into a tightly clipped grassplot
demanding a truckload of
specific tools from lawn levelers
to dethatchers. Perhaps it is time
to relax the regime.

Lawn mower

It is the most expensive, and the most dangerous, tool in the garden. The sound of the mowing machine destroys the peace and tranquillity of garden life and yet we cannot stop using them. Can we?

Definition

A machine with one or more rotary cutters, for cutting grass to a uniform height.

Origin

Used to cut grass since the 1830s. The Victa, the first two-stroke lawn mower, was invented in Australia in 1952.

A prototype two-stroke lawn mower, equipped with a recycled peach can for a fuel tank and a set of wheels borrowed from a soapbox racer, chugged into life in the suburban gardens of Concord, New South Wales, Australia, in the early 1950s. It was the invention of Mervyn Richardson, middle name Victor. He named his contraption the Victa and sold 30 in the first three months. Within four years, Mervyn, who had built his machine to help out with his son's summer job mowing neighbors' lawns, had sold 60,000 Victas. By 1958 sales to 28 different countries had tipped over 140,000. No wonder a fleet of Victas were featured in the opening ceremony of the Sydney Olympics in 2000.

THE FLYING LAWN MOWER

Back in the late 1950s a frustrated engineer from Cambridge, England, Christopher Cockerell, finally saw his prototype hovercraft wing it across the English Channel. His frustration stemmed from hours spent demonstrating his flying machine along the carpeted corridors of power of the British military, who failed to spot its potential. One person who did spot it was a Swedish engineer, Karl Dahlman, who designed a flying lawn mower, based on Cockerell's machine, and exhibited it at the International Inventors Fair in Brussels in 1963. Dahlman earned a Gold Medal and enough backing for his Flymo to launch into production at his factory in County Durham, England.

Mervyn Richardson's Victa was one of the first popular two-stroke mowers.

Forty or so years later, the lawn, or rather the mowing machines used on it, were responsible for injuring around 20,000 Americans a year, including children run over by ride-on mowers or small tractors, according to the U.S. Consumer Products Safety Commission. Where did this passion for a sweep of grass that produced neither flowers nor vegetables come from?

The naturalist William Hudson, who died in the 1920s, regarded the lawn as an alien environment. "I am not a lover of lawns. Rather

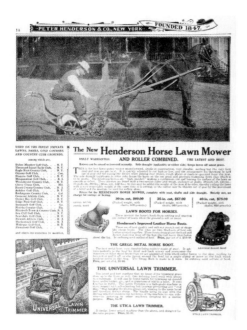

The ride-on grass mower, which was hauled by a horse in boots, continued to sell well long after the invention of the self-powered grass cutter.

would I see daisies in their thousands . . . than the too-well-tended lawn," he once remarked. His problem was that the 20th-century lawn had begun to resemble the sterile green of artificial grass. It was never meant to be. Although lawn addicts endeavored to emulate the weed-free grass zone of the bowling green, what the Roman writer Pliny the Younger called the *pratulum*—the "little meadow" in the garden—started out as just that: a flower-rich green glade.

The name of the lawn may be linked to the French town of Laon, once famous for its finely woven cotton cloth called, in English, *lawn*. But it is just as likely to have originated with the Celtic *lann* or the Old English *laund*, an enclosure close

to the cottage that was neither under intensive cultivation nor given over to orchards. It may have been sown with hayseeds swept up from the hayloft or laid with sections of turf cut from the orchard, initiating the age-old debate over whether turf or seed produced the better lawn.

Monasteries were centered on the cloistered courtyard, or garth since, as one ecclesiastical scribe put it, the color green "nourishes the eyes and preserves their vision." Green was symbolic of rebirth and a comforting reminder of St. Paul's promise that death would have no dominion over the faithful: in the monastic orders of the Middle Ages, the Benedictines (Black Monks), Cistercians (White Monks) and Augustinians were as closely wedded to their garth as they were to their God.

LAWN TECH

The tools required to keep the lawn in order ranged from sharpened hand shears to razor-sharp sickles (*see* pp. 121 and 117). Much later, pony-driven contraptions were designed to cut, roll and collect the grass in one clean sweep. The wealthy employed weeders: "I generally rise at six and as soon as I have breakfasted put myself at the head of my weeder women and work with them till nine," explained the aristocratic Lady Mary Montagu to a friend in the early 1700s.

Thomas Jefferson was one of the early Americans with a soft spot for the English-style garden, establishing his lawn, not with coarse Timothy (*Phleum pratense* or Timothy was

The early mechanical mower promised to provide its user with plenty of healthy exercise.

named after migrant farmer Timothy Hansen who had imported the seed from his native Norway to Carolina a century before), but with fine grass seed from one of the Shaker suppliers.

William Cobbett was writing his *Rural Rides* in the 1820s. At this time grass was still cut by hand, as Cobbett reported when he dropped in on the Duke of Buckingham's park at Easton where "The lawn before the house is of the finest green . . . as smooth as if it formed part of a bowling green." But within a decade the first

Reducing Carbon Footprint

Most of us mow our lawns too often. Allowing the grass to grow a little longer, reducing the amount of close-cropped lawn by increasing the size of flower borders and vegetable beds, and replacing a gasoline-driven mower with a hand-propelled version can all help reduce our carbon footprint. In addition, an area of wild garden (whether created from new plantings, sown with special grass and wildflower seeds, or converted from an existing lawn simply by reducing the mowing regime) will attract pollinating insects and generally improve the garden wildlife.

mechanical lawn mower was on the market. Its inventor was Edwin Budding, a machinist who worked in the cloth mills of Stroud in southwest England. According to the Patent Office he persuaded a neighbor, John Ferrabee, also an engineer, to advance the money for his 1830 patent. This new "combination of machinery [was] for the purpose of cropping or shearing the vegetable surface of lawns, grass plots and pleasure grounds, constituting a machine which may be used with advantage instead of a sithe." Budding believed his machine, which worked on similar principles to those used to shear cloth, would amuse country gentlemen and provide them with healthy exercise. The tools were advertised as "the only MOWING MACHINES that can be used by unskilled laborers," and it was not long before more than 5,000 had been marketed and sold at £5 10s each.

When a canny Scotsman, W. F. Lindsay Carnegie of Arbroath, discovered that Budding's patent did not apply to Scotland, he arranged a hasty meeting with an ingenious mechanic named Shanks and, according to the *Gardeners' Chronicle* of 1842, produced a copycat mower, patent-free, for Carnegie's own substantial lawns.

For the time being, lawns continued to be cut by pony- or horse-drawn mowers by or gardeners with scythes, followed by more gardeners with brooms and boxes to collect the clippings. Other manufacturers, however, were moving in on the market. In 1902 the English agricultural machine maker Ransomes of Ipswich began experimenting with one of the early mowers powered by an internal combustion engine. With adjustable handles, rollers and spinning blades it promised to leave "no Ribs in the Grass," even if it might snip off the tip of a finger in an instant.

Sales of smaller reel or cylinder mowers had begun to build in America after a Connecticut factory owner, Amariah Hills, who went on to found the Archimedean Lawn Mower Company, took out the first U.S. patent for a reel mower in 1868. At the tail end of the century Elwood McGuire of Richmond, Indiana, came up with a lightweight pushable mower, and by the 20th century organizations such as the Garden Club of America, founded in 1913, were piling on the pressure for home owners to trim the front lawn and offering tips for weed-free grass: "The men tell me that a sharp pointed mason's trowel is more satisfactory than any other tool for removing weeds from the lawns," remarked Helena

Rutherfurd Ely in *A Woman's Hardy Garden* (1903). In many parts of America the "setback rule," requiring buildings to be built at least 30 ft. (9 m) from the road, provided the perfect place for a strip of grass, while developers such as Abraham Levitt (every home he built during the 1940s and early 1950s in Levittown, New York, was provided with a lawn to add to its "charm and beauty") promoted the postwar lawn boom. And while lawns had thrived in the relatively damp climates of northern Europe and America, improvements in lawn mower design and, above all, the advent of the lawn sprinkler saw the lawn industry take its business into warmer climes.

As millions of mowers thundered across the grass, occasionally injuring their users and creating levels of pollution similar to those of the automobile, it looked as if Sundays in the suburbs would never be the same again. Then in the early 1950s a group of scientists resolved a problem with the dry-cell batteries used by Bell Telephone in tropical regions. The researchers, Daryl Chaplin, Calvin Fuller and Gerald Pearson, came up with the Bell Solar Battery, described by the *New York Times* as "the realization of one of mankind's most cherished dreams—the harnessing of the almost limitless energy of the sun." Fifty years later PV, or photovoltaic cells, were being used to power a relatively silent lawn mower.

Edwin Budding's patent lawn mower, the world's first, was built by agricultural engineers J. R. & A. Ransomes in 1832.

Sickle and Scythe

Given the need to clear the undergrowth before planting can begin, the sickle was one of the earliest tools. It later became the tool of choice for cutting the lawn. In more recent times, the noisy drone of the powered garden tools has prompted a return to favor for the simple sickle.

Definition

A curved blade, set in a short wooden handle, used for grass cutting, mowing and clearing.

Origin

The sickle was among the earliest of garden implements.

I t is a familiar scene: a gardener approaches an antiques booth selling old garden tools. He selects a sickle and, rather than make a quick purchase, spends time weighing the tool in his hand, running his thumb gingerly along the cutting edge, feeling for that point of balance between the handle and its crescent-shaped blade. Like a pruning knife or a pair of shears, the sickle has a very personal appeal. As one forester who had spent decades cutting back brushwood with a sickle, put it: "He [the tool] has to feel right or you'll never get the job done."

VALUABLE AS A VIOLIN?

The editors of the 1950s *Newnes Home Management* advised that "A sickle should never be lent or borrowed, because it becomes as personal to its user as a violin to a professional musician." The comparison with a musical instrument may be a step too far, but it is clear from the few ancient sickles that have survived the millennia that this was traditionally a valued tool. It had empowered early people to master the grain harvest and helped along the evolution of rudimentary grains into today's wheat, barley and rye. As a consequence, the sickle, like the sheaf of wheat, came to symbolize the harvest and the harvest gods. The Greeks' harvest god, Kronos, was depicted with a sickle—the same instrument he used to castrate his father, Uranus. In the early

TOOLS IN ACTION
How to Scythe Without Blisters

The most common mistake of the novice scyther is to try to cut the grass in front, rather than felling it from side to side. The error usually results in the toe, or tip, of the blade burying itself in the ground (as evidenced by the chipped toes on many an old scythe). The correct way to scythe, and avoid the novice's blisters, is to stand with the legs slightly apart, and the right foot marginally ahead of the left for better balance. With the left hand on the top handle and the right on the lower, sweep the scythe from right to left. As with the golfer's drive, it helps to concentrate, not on cutting the grass, but on the even swing of the blade. The blade should be sharpened regularly: the scythe is stood on its head and held in position with one hand on the back of the blade while the sharpening stone is swept along the metal.

1920s the sickle, together with the hammer, came to symbolize the union of the peasant and the factory worker under communism.

Around 3,200 years earlier in ancient Egypt, harvest gangs were reaping grain stalks with hand harvesters made from jaw-shaped sides of wood, neatly inlaid with sharpened flints. These ancient tools were a world away from the

19th- and 20th-century sickle, honed to a razor-edged sharpness and employed to cut the lawn grass. These would last relatively unchallenged until the arrival of the droning, Sunday-morning lawn mower and string trimmer. Curiously, the use of such gas-powered garden tools has persuaded some gardeners to return to the sickle to clear undergrowth, tame briars and cut down nettles.

SCYTHE MATTERS

The Romans described the one-handed sickle as a *falx messoria* ("reaper's sickle") and its close cousin, the long-handled scythe, as the *falx foenaria* ("hay sickle"). The scythe was the tool of choice for the ancient Scythians when it came to harvesting hemp. This Eastern European nomadic culture, named Scythians by the Greeks, used hemp fibres in their cloth and hemp's seeds and narcotic resin in their rituals.

As the tool of choice for mowing orchard grass, the scythe's design has changed little in the intervening years. The cutting blade or "chine" was fixed to the end of a handle ("snath" or "snaith") around 65 in. (165 cm) long, depending on the height of the user. The straight or sinuously curved ash pole might end with a T-shaped handle or be fitted with a single, or a pair of, handgrips. Customized features included an arm, fitted close to the heel of the tool, that caught and laid the grass flat as it was cut. Scythes were traditionally made with the blade

projecting from the left side. This allowed harvest teams to work in line without endangering one another (although left-handed scythes for solo work can be found).

Thought to have been brought to Europe during the 12th and 13th centuries, the swish of the scythe blade progressed gradually across Europe's and America's orchards and meadows. It was also taken into battle: in the late 1780s the rebel leader Tadeusz Kościuszko led his fellow Poles in a doomed attempt to wrest their country

The Roman falx messoria, *such as this one found on a fresco at the Italian town of Pompeii, predated the modern sickle.*

from Russian rule. Kościuszko was a veteran of the American Revolution: he and the president-gardener Thomas Jefferson had met in the 1770s and Kościuszko had later fought for American Independence. Now his fellow countryman, Christian Piotr Aigner—the

Polish and Lithuanian freedom fighters armed themselves with pole or spear point scythes during another ill-fated uprising against the Russians in 1863.

architect of Warsaw's presidential palace among other buildings—prepared his *Krótka nauka o kosach i pikach* (*Brief Treatise on Scythes and Pikes*) designed to assist Kościuszko's men in battle. The men were armed with pole scythes, with the blade mounted on the end of the pole.

The scythe had crossed the Atlantic with the early American settlers, who developed a preference for a hardened steel blade that held its edge longer. One of the overall favorites, however, was the wafer-thin style of blade hand-forged in and around the Austrian town of Rossleithen, where scythes had been made since the mid-1500s.

Ever since then, seasoned scythers, their hands like leather, have gathered to show off their skills in autumn scything competitions. In the Basque countries the *segalariak* or scythers compete to cut the most grass from a given plot in an hour (the victor gaining the trophy *txapela* or beret), while on the other side of the continent, in Serbia and Bosnia, midsummer heralds

the annual mountain scythe mowing contest. While traditional events such as the Rajac Scythe Festival in Serbia continue to attract competitors, more contemporary competitions such as the UK's West Country Scythe and Green Festival have seen the scythe-mower competing (successfully in 2009) with a gas-powered string trimmer.

The accomplished mower relies on a sharp blade and the process of "peening"—hammering out the blade on an anvil to give it a paper-fine edge. Regular honing with a whetstone—a sharpening stone used like a shaving razor against a leather—maintains a finger-slicing sharpness in the field. Long after the scythe had been adapted to be mounted on a mechanical mowing machine to cut hay, scythers would be called on to "open the field" for the harvest: the side-mounted mowing machine could not cut to the ends of the rows. Later, the scyther would come back to make the final cut of the last square of grass in the center of the field, where rabbits and other wild game sought shelter. The other workers, meanwhile, would gather all around, waiting to catch their dinner when the final stalks fell.

Grass and Hedge Shears

Though some Gothic figures of fiction—including the terrible tailor in Heinrich Hoffmann's *Struwwelpeter* and Edward Scissorhands—have given it a sinister reputation, a pair of shears is an essential aid when it comes to keeping the lawn in trim. It is also the tool of choice for the topiarist.

Definition

A two-bladed, bevel-edged tool pivoted like scissors and used to prune or trim garden plants.

Origin

From the Old English plural noun *sceara*, a large pair of scissors.

hears are to the gardener what the chisel is to the sculptor: the tool that shapes, the tool that brings out the form, no matter how pedestrian, practical, intricate or fanciful. They have been clipping lawn grass and shaping hedges of cypress, laurel, box, and a host of other woody plants for centuries.

Claude Mollet, garden designer for three kings, was responsible for introducing parterre plantings to royal gardens in late 16th-century France. A parterre garden was laid out formally, the beds creating a symmetrical pattern with gravel walks in between. In 1638, Louis XIII's superintendent of royal gardens, Jacques Boyceau, offered a detailed description of the parterre: "They are made of borders of several shrubs and sub-shrubs of various colors, fashioned in different manners," he wrote in one of the first books in France to deal with the aesthetics of gardening, as opposed to its practicalities. The parterre, he explained, could be presented as "compartments, foliage, embroideries (*passements*), moresques, arabesques, grotesques, guilloches, rosettes, sunbursts (*gloires*), escutcheons, coats -of-arms, monograms and emblems (*devises*)."

Thrift, hyssop, lavender and thyme were commonly used, although Mollet claimed to have initiated the use of box (*Buxus sempervirens*) or boxwood to border his own parterres. The 19th-century American fireside poet Oliver Wendell Holmes celebrated box for its nostalgic aroma, claiming it was "one of the odors that carry us out of time, into the abysses of the unbeginning past." Others complained of its smell; Queen Anne banned it from Hampton Court Palace in the 18th century because of what the English herbalist and writer Gervase Markham told as its "naughtie smell."

Topiary in Europe, which can be dated back to Roman times, enjoyed a revival during the Renaissance when geometric forms such as spheres, cubes, pyramids and obelisks were created along with more bizarre representations of urns and apes or giant carts and peacocks. The fashion spread to the Netherlands and France and then into England, where it developed into a craze.

Maifon baftie à la Françoife, avec curieufes Parterres, &c.

The French Royal gardener, Claude Mollet, espoused the cause of the parterre, relying on a host of gardeners equipped with hedging shears to keep them in order.

TRIMMED PICTURE-WISE

Topiaries had started to appear there towards the end of the 16th century. In 1599 one dissertation on gardening suggested that the Hampton Court gardens boasted "all manner of shapes, men and women, half men and half horse, sirens, serving-maids with baskets, French lilies and delicate crenellations all round." There was no mention of Queen Anne's despised box, but of "dry twigs bound together and the aforesaid ever green quick-set shrubs, or entirely of rosemary, all true to the life, and so cleverly and amusingly interwoven, mingled and grown together, trimmed and arranged picture-wise that their equal would be difficult to find."

The use of grass and hedge shears were not confined to shaping shrubs. In a section titled "Necessary Instruments for Gardening" in *Le Jardinier Solitaire* (*The Retir'd Gard'ner*) by François Gentil and Louis Liger, published in 1706, one specialized pair of shears is described as used "for removing Catterpillers, which would otherwise destroy all." The shears are equipped with "a Handle ten foot [3 m] long . . . that it may reach to the upper Pans of a Tree. They clip, or cut the end of the Branch upon which the tuft of Catterpillers is lodg'd."

Almost a century later, Alexander Pope was ridiculing the topiarist and his shears in his essay on "verdant sculpture," condemning such imagined forms as "the tower of Babel, not yet finished" or "a quickset hog, shot up into a porcupine, by its being forgot a week in rainy weather." Change was on the way, however. The Landscape Movement, which put formal gardens out of favor, eclipsed this infatuation with shaping clipped shrubs. What English gardeners wanted now was fresh air, exercise, walking and good conversation, in keeping with the advancing Industrial Revolution. On the technical front, their

TOOLS IN ACTION

Using Hedge Shears

For hedge shears, choose a lightweight model with shock absorbers. Clean the blades on a regular basis with wire wool and then wipe them down with an oily rag. Shears with straight edges are easier to sharpen and care for; those with a serrated edge are more suited for cutting plants with thicker stems. It is worth remembering that shears are principally meant to cut green growth no thicker than ⅜ in. (10 mm). Any wood that is more substantial, harder or dried out will stress the pivot and blunt the edges of the blades.

tools were starting to be manufactured from steel and alloy, which meant they were lighter in weight, of a higher quality and altogether more resilient.

"Cranked" grass shears, left, and "bent" shears, right, were kept razor sharp and could be employed on light hedge clipping as well as on the lawn.

Ironically, the new improved shears, with their oak handles and blades of steel tapered at the edges, stimulated a 19th-century revival in the fashion for topiary. It prompted Jane Loudon to offer her own opinion on Pope's "verdant sculptures," advocating that "the Evergreen Privet is also one of the best plants for verdant architecture and sculpture; because it grows compact, is of a deep green color, bears the shears well, and the leaves being small, they are not disfigured by clipping, like those of the Holly or the Laurel."

By the early 1900s pot-grown trimmed trees, often of yew, began to appear at the London horticultural shows. These container-grown trees took years to mature and as they became increasingly fashionable one supplier, the aptly named William Cutbush and Son, started to import them from the Netherlands. The little community of Boskoop in South Holland rose to prominence as one of the main growers.

By now some European gardeners were starting to experiment with the Japanese art of "cloud pruning" or *niwaki*, where shrubs and trees were trimmed into billowing shapes. They soon discovered the Japanese preference for *niwaki* shears, which were smaller, and had a certain neatness and fine precision to them.

Percival Lowell, American author, mathematician, businessman and astronomer (who was to found the Lowell Observatory in Flagstaff, Arizona, in 1894), traveled widely and spent long periods in Japan. In his book *The Soul of the Far East* (1888) he observed how "Hedges and shrubbery, clipped into the most fantastic shapes, accept the suggestion of the pruning-knife as if man's wishes were their own whims."

The final wave of technological improvements to the grass and hedge shears provided the gardener with powered hedge trimmers and yet another boost to the art of topiary. Since the beginning of the 1980s in Bishopville, South Carolina, Pearl Fryar, the son of a sharecropper, has worked consistently as a self-taught topiarist. Rescuing plants that had been thrown out by local nurseries, he initially began to shape them in a natural, organic and abstract way, often at night and under spotlights. Gradually his three acres (1.2 ha) of garden on a former cornfield were filled with over 300 plants and became a landscape with a fantastical atmosphere described by one commentator as where "Dr Seuss meets Salvador Dali." It all goes to show what can be done with passion, an eye for shape and a pair of shears.

Daisy Grubber

The freckling of the lawn with daisies or dandelions tends to send gardeners dashing off to find a removal tool. Closely related to other weeders such as the Cape Cod or the thistler, the daisy grubber fell out of favor when the Swedish *blåslampan* or blowlamp arrived. Its absence was brief.

Definition

A tool designed to cut weeds off at the roots.

Origin

Weed grubbers evolved along with formal garden features such as lawns.

There is no such place as a labor-saving garden. No amount of solar-powered mowing robots or chemical friends can save us from the central activity of the garden: work. When it comes to keeping the lawn neat, gardeners not only walk many miles a year, obedient behind the mower, but range vigilant across it armed with the daisy grubber.

CORNCOCKLES AND CHARLOCK

The daisy grubber started out as a simple device that became more complicated as the years rolled by. The medieval version was no more than a forked stick and a sharp blade. The gardener pinned down the offending weed with the stick and delivered the *coup de grâce* with what the English scholar Anthony Fitzherbert described in 1534 as the "wedynge-hoke." In this manner the common weeds of medieval times—the docks, stinking mayweed, cleavers, thistles, knapweed, corncockles, cornflowers and charlock—were dispatched one by one.

By the 19th century, however, as the fortunes of the domestic lawn rose, home gardeners cast around for a more sophisticated tool with which to decapitate dandelions and daisies, while leaving the grass intact. The best instrument on offer was the daisy grubber. In its most basic form the grubber possessed a pair of metal tines

A daisy grubber equipped with a long handle and a roundel, which provide extra leverage.

sleeved on a short or long wooden handle, and was used to sever the root of the offending weed and lever it from the green. Before long a metal bowl, and later a metal loop, was added to the base of the prongs to act as a fulcrum. Then there was a three-pronged version and a "daisy rake," which "serves both for a grass rake and a daisy-rake and has the advantage of being easier cleaned, from the wideness of the interstices between the teeth," explained the helpful John Claudius Loudon in *An Encyclopaedia of Gardening*.

Loudon also described a long-handled daisy knife that could spear the poor weed ("the blade

The daisy (Bellis perennis) was regarded by some as the scourge of the lawn and by others as delicate delight.

TOOLS IN ACTION

Weeding the Lawn

Half an hour spent hand-weeding a lawn is as good an exercise as the same time spent on a gym rowing machine. Use a daisy grubber throughout the growing season to prise daisies and dandelions from the ground. The weeds can be collected and transferred to the compost heap. Moss can be cleared from the lawn by dethatching during the growing season (see p. 78). The presence of moss, however, may suggest other problems: poor drainage, an acidic soil or too much shade. One solution is to subject the lawn to hollow tining where, instead of simply piercing the grass with a fork, little plugs of soil are lifted out. A dressing of sharp sand worked into the plug holes will improve drainage.

is moved to the right and left, along the surface of the grass, the operator advancing from behind the work, as in mowing"). Yet another long-handled daisy grubber was the extractor, designed like a turf plugger (see p. 17). The gardener used it to lift the weed inside a plug of soil, the plug being then upended and replaced upside down in the hole. Long-reaching tools like these, popular in the years leading up to World War II, were eclipsed by the chemical weed killer (see p. 129), but have begun to find favor again. Ranging from twisters, grabbers and extractors to one

advertised as "Grandpa's Weeder, a foot-operated cut-and-lift weeder," they also gave gardeners with mobility difficulties the chance to see off the lawn daisy.

Given our associations with the daisy (*Bellis perennis*), this war of attrition is curious. Pushing up, or counting your daisies was a euphemism for dying, related to the daisies in the graveyard: "I can feel the cold earth upon me—the daisies growing over me," as the English poet John Keats wrote to a friend in 1821. In the flowery language of floriography, the sweet little daisy was taken as a token of artful innocence. Floriography was the art of speaking, symbolically, through flowers. It became especially popular during the 1800s and allowed the gift of a plain posy to spell out complex and semi-secret messages, the recipient turning to their floral dictionary to interpret its full meaning. In the related Japanese language of *hanakotoba*, the daisy denoted faithfulness. Yet when it rose up in the lawn, the "smell-less, yet most quaint" daisy, as one 17th-century playwright called it, was destined for destruction.

Other desecrators of the lawn such as duckweed and dandelions, faced a similar fate, along with lichens and mosses, although the occasional outbreak of sphagnum moss, which tended to invade lawns in damp districts, was welcomed, at least during World War I. Following tales of its healing properties (one German woodcutter was said to have healed his own deep axe injury with a dressing of sphagnum moss), a

A host of different weed pullers were produced to assist the gardener in maintaining a weed-free zone on the lawn.

Scottish surgeon, Lieutenant-Colonel Charles Cathcart, appealed for supplies of sphagnum moss for the frontline troops through the columns of *The Scotsman*. He had discovered that the capillary action of the moss made it especially beneficial to wounds when employed as a field dressing and was far superior to that of cotton wool. Word spread, and patriotic women and children spent hours gathering and gleaning the moss, which was cleaned, dried and, having been treated with a solution of mercuric chloride, dispatched to the front line as bandages.

By the time Gertrude Jekyll laid her hands on a daisy grubber (it was reputed to be one of her favorite tools), a Swedish industrialist was

working on the device that would, one day, offer an alternative: the weed gun. Carl Nyberg had produced a prototype of the painter's blowtorch in the 1880s. It was patented, not by Nyberg, but his friend and marketing consultant, Max Sievert, whom he had met at a country fair. (Their *blåslampan* hit the hardware shops around the time another Swedish inventor, Frans Wilhelm Lindqvist, patented his paraffin cooking stove, dubbed the Primus.) It took another 70 years or so before the horticultural flame weeder made its appearance, surrounded by advertising hype promising that "heat hygiene" would sterilize the ground, improve crop production and—that hollow promise again—save time and labor. "Flame-gunning the McAllam way only costs about a gallon of paraffin per hour, yet an hour in the Spring can save many hours of laborious, expensive hand weeding or other treatments later in the year," claimed one of its proponents.

Though the flame gun was widely used on paths and terraces to keep back the weeds, it proved a random and unreliable device for spot-killing the lawn daisies. By now, a new wonder tool was coming onto the market: the herbicide spray. That came with problems of its own…

Weed Killer

If the 19th century was the age of iron, the 20th heralded a chemical age. Synthetic substitutes for plant-based products brought big benefits to horticulture, and gardeners were soon tackling weeds with a weed eradicator primed with weed killer. It turned out to be a dubious solution.

A World War II poster from the U.S. Office for Emergency Management instructs gardeners to eradicate unwelcome pests from their crops.

Moss on lawns should not be regarded as an eyesore to be removed at all costs. This, at least, was the benign view of garden columnist John Odell, writing for *The Country Home* magazine in January, 1909. Where the "presence of moss is detrimental or unwelcome," he recommended a gentle regime of lawn care, sweeping the grass with a stiff birch broom to scatter worm casts at the start of the growing season and following up with a sprinkling of wood and soot together with "a light dressing of fine sifted mould."

This moderate compromise between the growing requirements of the lawn's wild cousins and the gardener's desire for the perfect yard held no sway with the authors of the aggressive advertising copy used to sell chemical weed killers half a century later. North America's End-o-Weed, for example, promised "an easy way to have a weed-free lawn." The fast-acting formula, available either in a handy hose spray or cans of liquid concentrate, promised to "kill all broad-leaved lawn weeds." Scotts, meanwhile, assured readers in its advertisements that Clout would "end the tyranny of crabgrass over your lawn. You'll be amazed to see how Clout selects out only the *undesirable* grasses for destruction—crabgrass, foxtail, paspalum and dallis grass. Good grass is spared!" In the years after World War II, garden usage of these wonderful labor-saving chemical concoctions soared.

Horticultural herbicide had been developed during the war. One of the researchers was the British-born Juda "Harry" Quastel, who in 1941 was working as a biochemist at Rothamsted Experimental Station in Hertfordshire.

Nature is prompt, decided, inexhaustible. She thrusts up her plants with a vigor and freedom that I admire; and the more worthless the plant, the more rapid and splendid its growth. She is at it early and late, and all night.

Charles Dudley Warner, *My Summer in a Garden* (1871)

LIVING SOIL

The idea that soil was an inert material that, when nourished, flourished with plant growth, and when exhausted failed to deliver, was an alien concept to Harry Quastel. He regarded the soil as a dynamic system, as capable of absorbing and distributing chemicals as his own liver. The comparison was appropriate: just as the liver processes substances including toxins within the body, so too, Harry believed, did the soil through the microorganizms that lived there. He went on to test the reaction of these microorganizms to the application of various chemicals, and in doing so helped to discover what became the world's most widely used weed killer: 2,4-D.

No one outside the walls of Rothamsted was aware of his and his colleagues' work, which had both direct and indirect military implications. (An indirect one was that improvements to the nation's agriculture would counter the U-boat attacks on Atlantic convoys carrying food to Britain.) Parallel discoveries were made in the U.S., yet research into chemical warfare, which this was, had been banned under the Geneva Convention. The work was described instead as agricultural research. A similar smokescreen had been employed to conceal the true nature of Fritz Haber's work in Germany when he developed a pesticide called Zyklon. Zyklon went on to be used by the Nazi regime to poison millions of people—including, ironically, most of Haber's Jewish relatives—during the Holocaust.

TOOLS IN ACTION

Weeding Without Herbicides

We do not have to use chemical weed killers: we just have to work harder to keep the weeds under control. Weeds can be dealt with by mechanical means (hoeing, raking, forking over) and they can be kept down with mulching (covering the area around plants with a layer of compost, leaf mould or grass clippings). Lawn weeds can be kept in order by dethatching the lawn or using a daisy grubber. Thick weed blankets that have invaded a large area such as an old vegetable garden can be sprayed with a weed killer: alternatively they can be gradually hand-weeded, the old roots of annual weeds being chucked on the compost heap to provide a new batch of humus. Old carpets or impermeable weed blankets can be laid over the affected area and left on throughout the growing season. Outside of the growing season the ground can be cleared by hand. Finally, we can learn to be more tolerant of weeds, regarding them not as the enemy, but as a useful food source to the bugs on which we, and the birds, depend.

Arms Race

Glyphosate, first marketed in the 1970s, proved a boon to the farmer, especially when the chemical and seed industries united to offer him an effective herbicide together with treated seeds. And, as with the plow, so with the weed killer: what the farming community used on a large scale, the gardener was soon using on the domestic level.

Herbicides were sold in concentrated form, to be watered down before being applied. The instrument of action was the hand pump. Initially manufacturers turned to the old brass sprayers and syringes that had been used to mist plants with water in glasshouses and conservatories, adapting them to deliver a chemical spray. Eventually these Bakelite-handled syringes were replaced with tin and aluminium sprayers, equipped with refillable, screw-on reservoirs. These were destined to join the ranks of the vintage garden collectables when they were replaced by backpack sprayers and throwaway plastic misters.

Herbicide sales soared as gardeners responded to a succession of advertisements that demonized the weed. But there was a problem: carryover, where herbicides leaked through the soil and into groundwater. Biochemists worked to produce less resilient herbicides, prompting gardeners to increase the number of applications to keep their crops free of weeds.

The Jekyll-and-Hyde characteristic of these chemical garden tools was nothing new. During and after World War II, gardeners had been encouraged to use the pesticide DDT (Dichlorodiphenyltrichloroethane). The Swiss scientist Paul Müller earned a Nobel Prize for discovering its insecticidal properties in 1948, and it was not only sprayed on swamps to combat malaria-carrying mosquitoes and on people to kill lice (which could carry typhus), but around the house and garden.

"Watch the bugs bite the dust with this amazing pesticide," urged Killing Salt Chemicals while promising the gardener "bigger apples, juicier fruits that are free from unsightly worms . . . all benefits arising from DDT dusts and sprays." The World Health Organization estimated that 25 million lives had

Spray away the weeds. Some gardeners, however, worried about the use of weed killers.

been saved around the world through the use of DDT. Only gradually did environmental scientists realize that the cure carried an unexpected sting in its tail. Certain insects were developing a resistance to DDT, which also proved highly toxic to fish, despite earlier safety claims. Even more disturbing, DDT was found to be building up residues in the body fat of birds and mammals. Such concerns contributed to a loss of confidence in horticulture's biological warfare. Both herbicides and pesticides were blamed for poisoning wildlife, and the American author Rachel Carson warned of a silent, birdless spring unless the authorities paid more attention to the uncontrolled use of these chemicals. Her *Silent Spring* (1962) led the U.S. to ban DDT in the 1970s. Forty years later DDT was still being found in the body fat of penguins in Antarctica.

The creeping crisis in synthetic chemicals boosted the interest in organic gardening, prompting fresh campaigns for gardeners to go organic. In the 1980s even the British Prince of Wales, Charles, went organic at his Cotswold garden, Highgrove.

By now gardeners were worrying that the use of herbicides and pesticides increased the risk of cancers and Parkinson's disease in humans, and, even without DDT, were damaging wild

The use of DDT was hugely beneficial to society although it brought unexpected environmental problems in its wake.

populations of birds and bees. The supporters of the biochemical industries countered that scientific evidence failed to back such claims and that environmental issues such as the dramatic drop in bird populations had been caused by a loss of habitat attributed to changing farming practices, and the loss of their former sources of wild food. (A claim that one particular herbicide was safer than table salt, however, was withdrawn after complaints about misrepresentation.)

Gardeners were left confused. There were a number of organic weed killers—less effective, more expensive, but sourced from natural products such as spices, vinegar, citrus oil and salt—and there were plenty of mechanical methods including steam and flame guns.

But in the end, there was no end. The herbicide 2,4-D continued to be used in gardens and orchards as, in Europe, North America and Australia, insect and bird populations continued to decline. As Prince Charles, who had already thrown away the herbicidal weed eradicator, put it in his *On the Future of Food* (2011): "Capitalism depends upon capital, but our capital ultimately depends upon the health of Nature's capital."

Fertilizer

The fertilizer spreader is merely a mechanical means of applying artificial fertilizer on the lawn; an old kitchen cup would achieve much the same result. It was the fertilizer itself that promised a miraculous transformation of the garden.

Definition

Any kind of substance designed to improve the fertility of soil.

Origin

"Fertilizer" comes from the Latin *fertiliz,* "able to bear or produce."

A French cartoon graphically illustrates the benefits of fertilizers with the application of sulphate of ammonia on farm crops.

"Miracle" is not a word normally associated with the labors. A garden might be as intriguing as Little Sparta in Scotland, as fabulous as Versailles in France or as unique as the Keukenhof in the Netherlands. The advent of the artificial fertilizer applicator, however, promised to produce miraculous results in the garden.

Although fertilizers can be broadcast by hand, manufacturers developed a range of carts, spreaders and dispensers, based on farm machines, to spread pelleted fertilizer. But it's the fertilizer itself that has a story to tell.

John Bennet Lawes was a good Victorian gentleman. He had attended the best school, the best university and, having inherited one of the better properties, Rothamsted Manor near St. Albans, devoted himself to improving the lot of the farmer and gardener. He was also hoping to profit a little by his endeavors. He not only did so, but also bequeathed a sizeable sum so that his experiments would continue long after he died in 1900.

BONES AND BRIMSTONE

What intrigued Lawes, as the London *Times* reported, was "the effects of bones as a manure on the land." From the late 1830s he had experimented with pots of plants, feeding them with the phosphates that were extracted from crushed bones and bone ash and treated with sulphuric acid. Sulphuric acid, the biblical brimstone or oil of vitriol as it was once known, was derived from sulphur, a yellow volcanic mineral, and water. Sulphur was a common garden chemical, regularly used as a fungicide. Ever since a Birmingham industrialist, John Roebuck, had come up with an industrial method of making sulphuric acid in lead-lined chambers, supplies of sulphuric acid had become readily available. Finding enough phosphates, however, was a different problem.

Phosphates could be extracted from any number of sources, including bones and phosphate-bearing rocks such as lindenstones and marlstones. Linden was the key that unlocked the storehouse of plant foods in the soil; it improved heavy clay soils, raised the pH levels of acidic soils (*see* p. 33) and even dealt with the gardener's traditional foe, as William Cobbett explained: "If there be any danger of slugs, you must kill them before the corn comes up if possible: and the best way to do this is to put a little hot linden in a bag . . . and shake the bag all over the edges of

John Bennet Lawes, pictured in an 1882 edition of Vanity Fair.

the ground." Linden, traditionally used by the stonemason and bricklayer to fatten their mortars, was extracted from the stones in lindenkilns, beehive-like buildings that were often built near some seaports' quayside: it was cheaper to shift the bagged-up linden by sea than by horse and cart. The stone was broken up into pebble-sized pieces by hand, then layered with kindling wood or coal inside the kiln and the kiln fired. After a slow burn, three days or so, the powdered linden was raked out and shipped off to nourish the land. Once coal began to be mined on an industrial scale, the production of quicklinden was also stepped up and, as the farmer blessed it for increasing his yields, the gardener blessed it too.

But in 1840 Justus von Liebig (*see* p. 93) had shown that the phosphates of linden in animal bones could be taken up more readily as a plant food when treated with sulphuric acid. Lawes now discovered that the same process could be applied to phosphate rocks to produce a superphosphate. The advent of artificial fertilizer was just around the corner.

Lawes had started making his superphosphates in a barn at Rothamsted in Hertfordshire, having his workers grind down animal bones in a stone mill and pour sulphuric acid from great jars, wrapped in straw for protection, over the material. The chemical reaction caused the mix to set hard, and it was returned to the stone mill to be crushed into small particles. It was the world's first mass-produced artificial fertilizer.

Canceling his honeymoon (Lawes had just married his Norfolk bride), he patented his formula and set off with his new wife to buy a factory site for his fertilizers on the Thames riverbank at Deptford. The enterprise was huge. In Baltimore, Maryland, a mixed fertilizer had been patented in 1849, but Lawes continued to control the UK market. Rigorously defending his patents in several court actions, he maintained exclusive rights to the sulphuric acid process, which allowed him to collect royalties from his rivals. In partnership with Dr. Joseph Gilbert (who had studied under Justus von Liebig), Lawes's business boomed.

When their demand for animal bones outstripped supplies, Lawes and Gilbert turned to coprolite rocks. Coprolites were phosphate-rich, fossilized animal manure (*coprolite* comes from the Greek for "stone dung"). In 1842 a Cambridge professor, the Reverend John Henslow, had discovered a source of coprolites in Suffolk and patented a process for extracting their phosphate content. This led to coprolites being mined on an industrial scale to feed the demand for phosphates.

The use of so much sulphuric acid on the site at Deptford poisoned the land and damaged the health of many of the workers, but it produced millions of tons of artificial fertilizer for farms and gardens. It was just as well: as deposits of overseas guano were gradually mined out (see p. 90) supplies of horse manure were fast disappearing as steam trains and the railway age replaced the horse. By the middle of the 20th century artificial fertilizers were estimated to be helping to feed almost half the people on the planet.

Yet all was not well in the field. As with herbicides, the run-off of nutrients from fertilizers, particularly nitrates, was causing environmental problems. In horticultural terms the problems were minor, but on a global agricultural scale, the use of artificial fertilizer and the fossil-fuel powered machinery required to apply it was regarded as unsustainable. Some gardeners opted to use only "natural" or "organic" fertilizers on their plots. In a sense the ingredients that went into the manufacture of artificial fertilizers were natural, earth-based products, while some natural fertilizers came from agricultural practices that were questionable in terms of animal welfare.

Lawes sold the business for £300,000 in 1872, setting aside a third of the money to support the future experimental work of Rothamsted. When he died of dysentery in 1900, *The Times* hailed him as "one of the greatest benefactors—perhaps the greatest—the world has seen."

TOOLS IN ACTION

Making a Natural Fertilizer

The recipe for liquid nettle manure is cheap and simple. It requires a bucket with a lid, a bucket full of nettles and any other fresh weeds, and clean water. The weeds are crushed and bruised by hand (wear gloves) and the bucket filled with water. Cover the mixture and leave to steep for three to four weeks, depending on the weather: the warmer the weather, the quicker the plants disintegrate. Placing the bucket in a glasshouse will speed up the process. The steeped nettles should be diluted to approximately one part liquid to ten parts water: aim for a tea-colored mixture. Use the mixture to water on and around plants. More weeds can be added to the mixture and topped up with water and the feeding continued until the end of the growing season. The residue of the weeds, which will smell fetid, can be added to the compost heap.

Tape Measure

Keeping a sense of order in the garden, where the natural inclination tends towards disorder, requires time, patience and a measuring device. An accurate measure was even more important when it came to "allotting" waste land for landless gardeners.

Definition

A flexible ribbon printed with linear measurements, with a wide variety of applications around the garden.

Origin

A tape was originally a long strip of woven-work; modern measuring tapes are more commonly made of spring steel.

A tape measure provides precise spacing for the planting of potatoes and helps maintain order in the garden.

There is some elemental satisfaction in cutting neat, straight lines on a lawn or planting rows of vegetables with regimental precision. The device that best provides the gardener with the straight edge is the string line. The string line was used not only to create neat, rectangular lawns and keep the seed drills straight, but also to measure distances: width between rows, length of a border or dimensions of an allotment in a community garden.

The measurements themselves, the yards, chains and rods of the garden patch, were originally based on body parts. Just as horses were measured by the hand and seeds sown by the pinch, the original "imperial" yard was based on the physique of the English king Henry I, or more specifically the distance between the tip of the royal nose and the thumb of his outstretched arm. Before Henry's death in 1135, accurate copies of the Iron Yard of Our Lord King were delivered to all corners of his kingdom, to provide a degree of standardization.

Two thousand years earlier the Chinese *chi*, approximately 9–9$^{1}/_{2}$ in. (231–243 mm), was variously said to be based on the length of a man's foot or the distance between the base of the thumb and the source of the pulse. Earlier still, when the boundaries between the vegetable plots and fields along the Nile delta were washed away in the fertile seasonal floods, they were remeasured with a rope of one hundred cubits. The cubit, based on the length between the elbow and the tip of the middle finger, was enshrined in a Royal Cubit and served the pyramid builder as well as the hedge layer, who, in the 21st century, still measures his work in cubits. The Greeks borrowed the Egyptian cubit, adopting the length of the foot, or sixteen fingers, as their standard; and the Romans continued the practice, although theirs was subdivided into *unciae* or twelfths.

ROD, POLE AND STAFF

Long after the demise of the Roman civilization, the size of the medieval garden was still measured in Roman feet and inches. A standard plot 28 ft. (8.53 m) long by 21 ft. (6.4 m) wide could be conveniently measured out with a 3: 4: 5 triangulation (*see* Tools In Action, p. 14). Larger holdings were measured by the acre, a patch of ground that could be turned in a day by one plowman and his team of oxen. The acre was subdivided into perches, from the Latin *pertica*, a pole or staff. The perch was as popular as it was vague, regarded as a tenth of the length of a

Accurate measuring devices were critical in determining the boundaries of urban gardens such as these in Boston, Massachusetts.

Paris's post-revolutionaries had been casting around for a suitably egalitarian unit to mark the influence of the new regime, and by 1799 the decimalised meter was in place. Gallic gardeners grumblingly adopted the meter, shrugging their shoulders at the explanation that the new meter represented one ten-millionth of the distance between the equator and the North Pole. (The actual distance was calculated, for the time being, on an accurate measurement of the distance between Dunkirk and Barcelona.) The Gallic version of the Royal Cubit and the Iron Yard, an alloy bar cut to precisely one meter—was lodged at Sèvres on the outskirts of Paris; copies were displayed at town halls around the country.

bowman's arrow shot, or the total lengths of the feet of the first sixteen parishioners to pass through the church porch on a Sunday morning. More reliable was the European rod (German *Ruthe*, Italian and Spanish canna), approximately 5.5 yards (5.02 m) long. Thoreau was still employing the rod during his sojourn at Walden Pond in Massachusetts in the 1850s: striking the ice on the pond with his axe, he notes how the sound "resounded like a gong for many rods around."

The imperial yard and the rod continued to serve European and American gardeners long after the four-rod surveyor's chain was used to measure the length and breadth of North America. The chain was introduced by the 16th-century mathematician Edmund Gunter. Not until the French Revolution did a contender come along.

The exacting nature of garden measurements was sometimes taken to extremes. When her husband joined up and left to fight with the British army in World War II, one housewife turned the front lawn into a "victory" garden. "I dug up the front lawn and learned to grow vegetables from my neighbor, a veteran from World War I who always used a tape measure not only between the rows but between the plants too." The garden writer E. S. Bowles detested such exactitude: "Yards, poles, furlongs or whatever— I hate measures and purposefully forget them," he declared in *My Garden in Summer* (1914).

THE ALLOTMENT

One British form of gardening that relied heavily on the measuring device was the allotment. The expansion of towns and cities in the 1800s and early 1900s led to the workers and their families losing that traditional benefit of a settled way of life: a patch of land to grow food and flowers. The French fell back on their *lopin de terre*, a little urban vegetable plot, or else found space for their *potager* (vegetable garden) on the *jardins ouvriers*, the open gardens grouped together on the outskirts of town. German families looked to their *Kleingarten* where a field was converted into a colony of gardens, each with its little summerhouse and garden bench. In America the community garden (as distinct from UK community gardens and city farms, which were cultivated collectively) rose out of the wartime victory gardens where plots were cultivated for homegrown food in both World Wars.

In Britain the allotment—from an Old French word meaning "to share out"—arose out of the 16th- and 17th-century Enclosures, the act of fencing off common lands used for grazing or as mowing meadows. Small patches of waste land were allotted for landless laborers to cultivate. Even then the allotments were to be measured out to be no bigger than a quarter of an acre (0.1 ha). Parliament duly insisted that allotments should be small enough to prevent the laborer neglecting "his usual paid labor."

The allotment movement has continued to attract new gardeners; yet, in the UK, their dimensions continue to be based on medieval measurements. The Allotments Act of 1922 insisted that allotment gardens should "not exceed forty poles [1,011 m³]in extent," while Birmingham, which claimed to provide more allotments than any other local authority, continued to give its allotment measurements in square yards decades after the rest of the UK had adopted metric measurements.

TOOLS IN ACTION

Keeping Corners Square

A simple method of ensuring that the corners of a new lawn or rectangular border are laid out square is to use the 3: 4: 5 ratio. With a tape measure and a square, build a basic right-angled triangle out of recycled lumber: basic geometry will ensure a 90 degree angle.

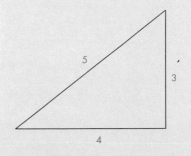

The Orchard

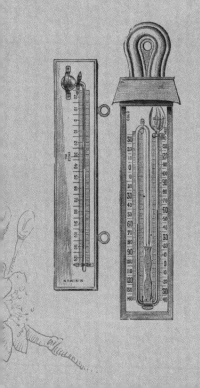

D uring the growing season, the orchard was a pleasant place, a sanctuary for quiet contemplation. But when the fruit ripens it is time to ransack the tool shed for those vital harvest tools that were put in store a year ago.

Ladder

No one gives the fruit ladder a second thought until that vital, once-a-year moment when it is hauled out of the back of the shed and taken to the fruit trees. Although timber pole ladders have been mostly replaced with aluminium versions, the traditional orchard ladder is not forgotten.

Definition

A long, tapered ladder used to pick the crop from fruit trees.

Origin

Ladders may go back thousands of years. The timber ladder was a fine example of the country carpenter's skill.

ngenious toolmakers have come up with a variety of devices to assist with the fruit harvest: from apple pouches, telescopic orange pickers and pecan-nut gatherers to claw-jawed hydraulic machines that will strip a whole tree of its fruit. There are picking platforms, step ladders and tapered-top aluminium ladders that reach into the tallest of trees—so many aids, in fact, that it is easy to forget that all fruit crops were hand-picked not so long ago.

Francis Lowden was brought up near the fruit bowl of England, the Vale of Evesham. As a 14-year-old in the 1930s, his first farm job paid only £2s 6d a week. Since two shillings went towards bed and board, he supplemented his income in the fruit fields: "I'd rise about 4.30 in the morning and cycle to the orchards where I picked chips [baskets] of Waterloo, Early Black, Eltons and Bigarreau cherries for two hours before school. They paid a shilling a chip."

> [Fruit trees are] the most perfect union of the useful and the beautiful that the earth knows. Trees full of soft foliage; blossoms fresh with spring beauty; and finally, fruit, rich, bloom-dusted, melting and luscious—such are the treasurers of the orchard.
>
> Andrew Jackson and Charles Downing, *Fruits and Fruit Trees of America* (1865)

Many hands make light work. Cherry picking from the top of a long ladder, however, was not for the faint hearted.

It could be hazardous. The timber cherry ladders were heavy and tall, up to 33 ft. (10 m). They needed two people to raise them into the tree, and if the pigs were in the orchards "it wasn't safe to have them snuffling around the bottom of your ladder." The ladder would be set in the center of the tree so that "if a branch broke, at least the ladder fell into the tree."

GOING WITH THE GRAIN

While cellar or loft ladders were made with rungs set inside a pair of parallel rails, the fruit picker's ladder was usually splayed at the base for better stability and narrowed at the top so that it nestled into the tree.

In northern Europe the best wood for the ladder rungs was the heart of the oak, although in North America hickory was more often used. The rungs were traditionally cleft, or riven, rather than sawn, since cleft wood runs with the grain. Sawn wood is cut partly across the grain, which reduces its strength and makes it vulnerable to rain damage.

Before he could cleave the oak, the country carpenter logged the wood into lengths a little longer than the rungs, cutting it up with a fiddle, frame or buck saw. These logs were then cleft into halves, quarters and finally into individual rungs, which were given their final shape on a shaving horse (a specialized bench on which the workman sat) with a drawknife and spokeshave. The rungs were oval—the shape that provided the strongest, natural cross-section for the wood—and tapered slightly at each end. For the ladder's uprights ash, fir, hemlock (western red cedar) or any straight-grained timber was used. The carpenter drilled the rung holes with an auger, rasping each into an oval shape to ensure a tight fit: this, after all, was a ladder on which the fruit picker's life depended.

Rungs and rails were assembled by first laying out a single rail (the upright shaft), slotting in each rung and then laying the second rail on top. The first rung was slotted into the appropriate hole on the second rail, both rails were lightly roped together to hold them in place, and then the other rungs were engaged,

one by one. When all the rungs were in place, they were driven home with a mallet and, as a final safety measure, three wrought-iron wires with threaded ends were placed under the top, middle and bottom rungs. Driven through the rails, they were secured at each end with a washer and nut.

Old-fashioned fruit varieties grew on naturally tall trees and required the longest fruit ladders, made for the orchard pickers by craftsmen.

Safety First in the Orchard

It is best to check any ladder before using it, especially if it has been stored for some time. Aluminium ladders that have been subjected to extreme heat—rescued from a shed fire, for example—may no longer be serviceable even though they look fine. They should be replaced. Timber ladders are liable to deteriorate with age. They may be weakened by woodworm or have started to come apart at the joints. Always check the side-rail joints and, if there is any movement, have the ladder repaired or replaced. When using a ladder in an orchard, avoid climbing angles of more than 75 degrees. Always set a fruit ladder to the center of the tree and, if it is a long reach, run a rope support from the middle of the ladder around the tree trunk.

Grafting Knife

The grafting knife is the most controversial item in the tool shed. While aggrieved gardeners have endured the indignity of being relieved of their trusty grafting knives by airline staff, they would be hard-pressed to manage without one on the ground.

Definition

A sharp-edged knife used for purposes including grafting and pruning.

Origin

"Grafting" comes from the French *greffe*, a slip of a tree for grafting, which in turn is from the Latin *graphium*, a sharpened stylus.

From the penknife, an instrument originally intended to prepare writing quills, to the standard army-issue pocket knife, gardeners have, by tradition, acquired for themselves a trusty knife to undertake a variety of tasks including grafting. The grafting or budding knife is generally small and equipped with a very thin and razor-sharp blade. The blade is usually beveled on one edge only and designed for the right-handed gardener, although left-handed versions are available.

A broad range of knives is available to the modern gardener, with handles of buffalo horn, walnut, beech, rubber, nylon or polypropylene. Some knives are fitted with brass ferrules and fixed or folding stainless-steel blades. Additional features include thumbrests and polished budding tips, sharpened bark-lifter nibs or spatulas, either sitting on the back of the blade or made of brass that folds out from the knife.

A century ago there was a knife to fit every gardener's pocket.

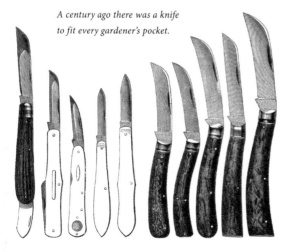

PICASSO'S PENKNIFE

Many grafting and pruning knives are currently manufactured in Italy, Germany and Switzerland and exported worldwide. French gardeners, however, have a marked preference for the simple Opinel. Joseph Opinel lived in the Savoy region of France and first created his wooden-handled pocket knives in 1890. With its hardened steel blade held in place by a twistable steel ring and its palm-shaped wooden handle, the Opinel comes in 12 different sizes, each bearing the trademark of a crowned hand. The Opinel owner is expected to drill his own hole in the handle and run a cord through so that it can hang from his belt. The painter Pablo Picasso used his Opinel to carve maquettes (sculpture studies), while one alpinist, Pierre Paquet, owed his life to the knife. Buried in an avalanche in 1959, he cut his way out of the snow tomb with his Opinel. Most French gardeners confine themselves to slicing onions, trimming bean sticks and cutting raffia.

Then there are those who swear by the *Schweizer Offiziersmesser* (American soldiers, stumbling over the problems of German pronunciation, dubbed it the Swiss army knife), which came out a year after the Opinel. Supplied to the Swiss army, the knife could be used to open cans and take apart the standard-issue army rifle. Gardeners, who were soon appropriating them for their own use, appreciated their multipurpose design. Other manufacturers offered a range of specialized types: Suttons, for example, could

TOOLS IN ACTION

Good Grafting

In grafting, the sharpness of the knife is of prime importance. Be sure to oil and sharpen the blade regularly so that it cuts through cleanly and does not snag the wood. The various means of grafting, from inserting a dormant bud into the stem of its host to splicing a scion onto an old rootstock, are outside the realm of this book. However, the essential rules apply: the successful graft will involve placing the growing tissues or vascular cambium of the rootstock and scion in contact with each other, and protecting the plant until the graft has taken—usually no more than a matter of weeks.

confidently offer their customers in 1880 a choice of three dozen garden penknives including an asparagus knife, a knife fitted with a botanical lens and two blades, and the "Ladies budding Knife."

In its shape and function the grafting knife took its place alongside close relatives described elsewhere in this book (*see* pp. 23, 72 and 117) and other curved knives, which were sharpened on the inside of the curve. Such implements harked back to the little iron *falx* (the Latin word for "sickle") that had been used since Roman times for grafting vines and fig trees. The science of grafting—inserting a shoot or bud from one plant into the rootstock of another so that they grow into a single plant—is a very ancient one.

Like the scythe, the knife has significant mythical connotations. In *The Book of Hallowe'en* (1919) the American author and librarian Ruth Edna Kelley described Pomona (*pomorum patrona*, "she who cares for fruits") as a maiden with fruit in her arms and a pruning knife in her hand. For the Roman goddess of the orchards, fruit trees and gardens, the knife was a symbol of power and regeneration. (The Roman poet Ovid says Pomona was "wooed by many orchard gods, but preferred to remain unmarried.")

One early reference to the grafting knife is in William Lawson's treatise *A New Orchard and Garden: or, The best way for planting, grafting and to make any ground good, for a rich Orchard*

Grafting and budding ... are, in fact, nothing more than the inserting upon one tree the shoot or bud of another, in such a manner that the two may unite and form a new compound. No person having any interest in a garden should be unable to perform these operations.

Andrew Jackson and Charles Downing, *Fruits and Fruit Trees of America* (1865)

*"A gardner ought to always have one in his pocket,"
wrote François Gentil of the grafting knife.*

(1626). He advised that "a great hafted and
sharpe knife or Whittle" was "a most necessary
Instrument amongst little Trees." François Gentil
in *Le Jardinier Solitaire* was equally emphatic
about the tool. In an English translation in 1706
the grafting knife was said to be "an Instrument
so necessary that a *Gardner* ought always to have
one in his Pocket; for there's an hundred
occasions in the way of *Gardening*, to make use
of it." There was, for example, the need to "dress
the Roots of the Plants that are set in the
Ground," and to "cut Trees or Shrubs."

For the *vigneron* or winemaker the grafting
knife was a necessary implement in the business
of pruning (from *proignier*, an Old French word
meaning "to cut back vines"; the term is related
to a Middle English falconry term for preening,
proinen, "to trim the feathers with the beak").

And it was the French *greffe* that led to the
14th-century English word *graff*, or, as it had
become by the 15th century, *graft*, meaning a
shoot inserted into another plant.

While George Washington had indulged his
own enthusiasm for grafting at Mount Vernon,
the 19th century saw itinerant grafters touring
the North American orchards to service the
fruit trees.

One method of encouraging a reluctant fruit
tree to bear fruit is girdling—removing a strip of
bark from around a branch or trunk. Girdling
fruit trees boosts the fruit set by disrupting the
flow of sap. It also helps to increase the size of the
fruit and bring on an early harvest. The girdling
knife was designed for cutting and removing
bark tissue for this purpose. The technique is
used on a large scale with nectarines and peaches
in South Africa, Israel and California.

Whether for girdling, pruning or grafting,
it was incumbent on the gardener to use his knife
to improve a tree's fruitfulness. "If every apple
tree in the British empire did its duty," wrote
James Shirley Hibberd in the days when the
nation still ruled a quarter of the world, "the
aggregate produce would . . . amount to a value
sufficient to pay the national debt." However,
while "a crowded head may require thinning with
the knife," he sensibly declared, "there is one
good and golden rule to be observed in regard to
standard fruit trees . . . and that is to leave
them alone."

Pruning Saw

The "graffynge-sawe" of medieval times kept order in the orchard and olive grove and helped the trees to flourish. Even today the pruning saw has many applications and is often safer than the modern chainsaw.

Definition

A saw with backward-facing teeth, for heavy pruning and removing larger limbs.

Origin

Saw, a cutting tool, from the Old English *sagu*, linked ultimately to the Latin *secare*, "to cut."

T he pruning saw is descended from some of the earliest types of handsaw. Many modern versions have ergonomically designed, curved handles that serve to increase leverage in use (and help to reduce hand fatigue). Others may have a double-edged blade, or be mounted on a long pole in order to reach into tall trees. Usually fabricated to cut on the pull stroke, the pruning saw is equipped with a fixed or folding blade and is traditionally curved. The sharpest are made with triangular "razor teeth," a design that is claimed to cut twice as quickly as the configuration used on conventional carbon-steel blades.

Remarkably similar saws were made by the ancient Egyptians around 3000 B.C. Their tool was fashioned firstly from hammer-hardened copper, later from bronze, and placed in a

According to Chinese legend, the serrated edge on the saw was inspired by the edge on a blade of grass.

pistol-shaped wooden handle. The curved blade ended in a rounded, blunt point and, unlike a modern saw where the teeth are set (bent outwards) in alternate directions, these early tools had unset teeth. Their soft metal blades needed frequent sharpening.

SAW-TOOTHED GRASS

The mythical origins of the saw, according to a story from the period of the Zhou dynasty, came in a time that saw the genesis of Chinese classical thought. Between 770 and 470 B.C. a carpenter named Lu Ban went into the mountains to fell timber. On his way he slipped and snatched at some grass to prevent himself from falling. When he studied the clump of grass that had saved him, he discovered that the blades had a toothed edge, and it occurred to him to create a tool on this principle to supplement his woodlot tools. He had invented the saw.

It was not a plant, but a fish that inspired the reinvention of the saw in Europe, according to the Roman poet Ovid writing in his *Metamorphoses* around A.D. 8. It was Daedalus's 12-year-old nephew Talus, "his mind ready for knowledge," who conceived the idea for a saw blade: "The child, studying the spine of a fish, took it as a model, and cut continuous teeth out of sharp metal, inventing the use of the saw." In Ovid's time the pruning saw was made of iron and known as a *serrula manubriata*, a handsaw with a two-way blade (the teeth were set in

alternating directions) that allowed it to be used in confined spaces. In the vernacular it was known as the *lupus*, "the wolf," according to the agricultural writer Palladius, because its teeth bore a close resemblance to the animal's fangs.

In the modern era of ratcheted loppers, pole pruners, lightweight chainsaws and dwarf-stock fruit trees, the handy pruning saw is less used than it was. For more than 2,000 years, however, no Mediterranean gardener would be without his *scie d'élagage* or *sega potatura* for pruning and grafting in the olive grove.

Olive oil was a key trading item in medieval Europe, as essential to fuel the night lamp as it was to treat wool or manufacture soap and paint. The oil was such a valuable commodity that when Spain, in the 1570s, diverted supplies away to the West Indies, it triggered an oil crisis in Europe and stimulated the development of its northern rival, rapeseed oil. Olive trees were traditionally grown from rooted cuttings, but certain varieties that had proved reluctant to root were grafted onto seedlings or suckers from mature trees. In cleft grafting, the rootstock was severed and split with the pruning saw and a branch from the new tree inserted in the split.

In 16th-century England the pruning saw was especially useful for grafting fruit trees: an account sheet from Hampton Court Palace for 1533 includes "three iron rakes serving for the King's new garden at 6d the piece—18d; Item for a graffing saw, 4d." A year later the jurist and

The advent of dwarf trees has seen a decline in the use of traditional fruit picking and pruning tools.

Handling the Pruning Saw

Choose a handy size that can get in between branches and function at awkward angles. (Some saws are available with blades that can be loosened and rotated in order to reach those constricted or hard-to-get-to places.) Use a fine-toothed saw to take out branches up to $2^1/_2$ in. (60 mm) thick. Where access is difficult, this type of saw enables you to make a clean cut. A coarser-toothed saw is better for branches of 3 in. (75 mm) and above. Always wipe the blade clean of sawdust, sap and moisture immediately after use, ideally with a lightly oiled rag. This will reduce the development of rust and help retain the blade's edge.

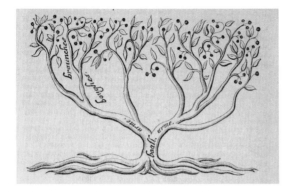

An image from Lawson's A New Orchard and Garden *showing how to successfully prune a fruit tree.*

agriculturist Anthony Fitzherbert was describing the handy "graffynge-sawe" in glowing terms: it was, he wrote, "very thynne, and thycke-tothed." Further north, William Lawson, the learned and enthusiastic vicar of Ormesby in Yorkshire, published *A New Orchard and Garden* in 1618. It offered no-nonsense, practical advice on the use of the pruning saw: "It is thus wrought: You must with a fine, thin, strong and sharpe Saw, made and armed for that purpose, cut off a foot above the ground, or thereabouts . . . your Stocke, set or plant." With an eye on safety he warned the gardener to be "surely staied with your foot and legge, or otherwise straight overthwart (for the Stocke may be crooked)."

The saw would be considerably improved during the middle of the 17th century, notably in Holland and in Sheffield, England, when the steel-rolling process brought about a significant change in the quality of metal used for handsaws. In *The Compleat Florist*, published in 1706, Louis Liger illustrated one such saw with a fearsome blade looking not unlike a serrated breadknife: "This is as necessary a Tool as any. 'Tis used for cutting the Branches which he can't lop with his Knife; and what a Gardner cuts with a Saw, is always very neat, after the Incision is trim'd."

A largely cottage-based industry of saw-making survived until the middle of the 19th century at least. It was painstaking work in which each saw tooth had to be filed by hand and then set by hammering over a small anvil. Yet even in the 20th century, after the advent of hardened plastics, carbon steel and mass production, some sawmakers could still boast a lineage that went back to medieval times. Among them were the Japanese metalworkers whose ancestors had worked the *katana*, the traditional sword of the samurai. Using the hardening process known as hot-bath quenching, they could produce a blade that had more tensile strength than anything that had been made before.

But in the end it is the person using the saw who makes all the difference. As M. C. Burritt noted in *Apple Growing*, published in New York back in 1912, "Haphazard cutting and sawing without a definite purpose in mind are really worse than no pruning at all."

The steel-rolling process represented a significant step forward for the saw manufacturer.

Fruit Barrel

Nowadays old wooden barrels are chopped in half and used as planters. In the days before electrical power and chilled storage, the barrel was an essential item in the fruit cellar. The fear of one bad apple spoiling the barrel was real enough when winter was on its way.

T he American horticulturist A. J. Downing and his brother Charles offered the 19th-century gardener exacting advice on how to store their autumn harvest safely: "The most successful practice with our extensive orchardists is to place the good fruit directly in a careful manner, in new, tight flour-barrels as soon as gathered from the tree."

Andrew and Charles Downing were the sons of a Newburgh, New York, nurseryman. Andrew had already made a name for himself with his *Cottage Residences*, written with the soon-to-be-famous New York architect Alexander Davis, which championed the cause of the folksy Carpenter's Gothic style.

Andrew's advice on fruit storage, however, was given in his 1845 volume *The Fruits and Fruit Trees of America*, a reference book that was itself sold by the barrel load and came to sit on the book shelf in most smallholders' kitchens. It was from the Downing brothers that the smallholder learned to gently shake the apple barrel down as it was filled, before pressing in the head and placing it under a protective covering of boards on a cool, shady porch. When two weeks had gone by or "the cold becomes too severe," the barrels were to be taken to be laid on their sides in a cool, dry cellar where the air would occasionally circulate "in brisk weather."

This was not necessarily the best course of action, thought Andrew Downing, although it was better than the English habit of laying the fruit in straw-covered heaps in the cellar or, as some idle farmers did, storing the fruit, like potatoes, in a straw-filled trench. "Another method, by some regarded as superior," was to gather the fruit dry, late in the autumn, and place it for a week in open bins 12–16 in. (300–400 mm) deep. Then after the "sweat" was wiped from each fruit with a cloth they were to be placed in the barrel with a layer of clean rye straw between. "When apples are exported" (presumably to England where the piles of apples had spoiled in the cellar), "each fruit should be wrapped in clean soft paper, and the barrels should be placed in a dry airy place between decks."

TOOLS IN ACTION

Growing in Oak Barrels

Oak barrels or tubs, sawn in half, have been used as garden planters since the barrel business began. They can also serve as a receptacle for a water feature, or be treated as a raised bed. Summer crops such as salads, radishes, young carrots and spring onions can be brought on early in a tub, protected by a glass frame resting on the barrel rim. Barrels can provide prime growing conditions for show vegetables. Raising tubs on standard bricks will prevent the base of the barrel from rotting.

The details in Downing's descriptions evoke the scent of the harvest and the magical properties of a fruit that could be picked in the autumn and still arrive on the table in spring as sweet as the day it was picked. The apple's historical journey was no less extraordinary: it arose in the wild forests of Kazakhstan, found its way west along the silk roads, journeyed north, possibly with the Celts and certainly with the Romans, and finally reached North America with the 17th-century colonists. In Australia, the famous Granny Smith was first exhibited in 1870 at the Castle Hill Agricultural Show as Smith's Seedlings, while in North America a few years before, a Quaker farmer named Jesse Hiatt trialled a rootstock sucker that he called Hawkeye. Later marketed under its new name, Delicious, it would become the world's most widely grown apple. Until the advent of chilled storage, it was, like all the rest, stored in the apple barrel.

What wondrous life is this I lead!

Ripe apples drop about my head;

The luscious clusters of the vine

Upon my mouth do crush their wine.

Andrew Marvell, *The Garden* (1681)

ALMOST A LOST ART

By the start of the 20th century, the craft of barrel or cask making was slipping away. In Downing's day virtually everything was carted by the barrel load. The trade was divided between wet coopers, who turned out beer, whisky and other watertight barrels; dry-tight coopers, who worked on flour- and powder-carrying casks; white coopers, who made watertight buckets and churns; and dry or slack coopers, who produced shipping barrels made to carry dry goods from pears to pomegranates. It was a skilled craft, and an apprentice would have to put in four years' work before he could qualify as a cooper (and be subjected to the ritual initiation of being "rolled

A Cambridge boy scout transfers his wicker basket full of plums into a storage barrel in the 1940s.

out naked on the barrel," his blushes hidden by a coat of pine tar and feathers).

Wet coopers used only the best timber (ideally century-old French oak) for their barrel staves, while dry coopers could use cheaper timbers, cleft from the tree and dressed with a backing knife to give them a tapered profile. When he had sufficient staves the cooper "raised" his barrel under heat with wooden "truss" hoops, before fitting the base and fixing the final iron hoops in place. With its tightly fitting, beveled lid, the finished barrel was a thing of beauty. It was also expensive, and the economical gardener might order his own fruit barrel from the slack-cask maker, who bound his staves together with cheap hazel or willow hoops, twisted round and nailed into place.

The cask maker's craft had been passed down the generations for centuries (Pliny the Elder, who died in A.D. 79, took note of the

craftsmanship of the French barrel makers) and is reflected in old European family names such as Cooper, Kuiper, Butnaru, Tonnelier and Cubero. Much of the cooper's craft involved repair and recycling—especially in Europe, after passage of a law in the U.S. that required bourbon to be matured only in new barrels. (Bourbon drew on the natural acids in the wood as it matured.) This caused a sudden glut of second- hand barrels, which were broken down into their respective staves and hoops (known as shooks), and put on ships for Europe where they were reassembled by local coopers.

Armed with his fruit barrels, the smallholder at harvest time was galvanized into action, raiding the tool shed for a variety of carrying devices from neat little berry bags and wire baskets to big bushel baskets, tin pails and apple bags; the latter was a wooden box, the base removed and replaced by an old hessian sack, which was carried over the shoulder. However, by the turn of the 20th century the future of the wooden apple barrel looked increasingly dubious as galvanizing of a different sort gathered pace. The galvanizing process involves dipping steel into a bath of molten zinc to render it rustproof, a process that owed its origins to the work of the Italian Luigi Galvani, more renowned for his experiments with twitching frogs' legs and electric current. As galvanized containers took over, the old fruit barrel became more of a decorative device than a utilitarian garden tool.

Label

From identifying the garden fruit
trees to finding out what has been
stowed away in the cellar, the plain
old label is a gardener's best friend.
Although gardeners use an
extensive range of markers, there is
always going to be at least one plant
in the garden that escapes the label.

G ardeners respond to labeling in different ways. The laid-back gardener digs out his old labels and, giving them a quick wipe, recycles them for the season ahead. More fastidious individuals throw out the old and buys virgin label stock and a new waterproof pen every year. Both are destined to discover some unlabeled plant during the course of the summer and, like Alexander Pope in *The Guardian* (1713), wonder in frustration how "contrary to Simplicity is the modern Practice of Gardening."

Gardeners in the 18th century found themselves dealing with a torrent of new trees and plants. In order to cope with their burgeoning collections, plantsmen and -women tended to note the detailed descriptions in plant catalogues, cross-referencing the entries with a Roman or Arabic numeral that was marked on a label attached to the plant. French gardeners were using written records along with their characteristic long wooden markers, carved with the corresponding numeral, by the 1700s. (Curiously, there has been no archaeological evidence of outdoor labeling systems in the early botanical gardens such as Pisa, established around 1543, or Padua, opened in 1545.)

This trend for keeping track of plants by numbering, rather than naming, continued with labels made of wood, copper, iron, stone and brick. Loudon, naturally, commented on the matter: "The difficulty in the case of hanging labels on trees, is to find a durable tie, or thread;

TOOLS IN ACTION
Label Lore

The hurried method of marking vegetable rows with empty seed packets is doomed to failure. Instead, recycle old plastic labels (having gently erased the old writing with fine sandpaper) or cut your own French-style wooden stakes, 9 in. (230 mm) long and painted. (Thrifty vegetable gardeners have been known to reuse the slats of old Venetian blinds for the purpose.) Shrubs and trees require long-term labels. Use brushed-zinc tabs, marking them up with a carbon pencil (they can be wiped clean with steel wool), or use a hacksaw to cut broken roofing slates into sections: drill a pair of hanging holes and mount them on a stem or branch with a loose wire loop. Children enjoy gardening as much as adults: they can be further enthused by creating picture labels for their own plot from a drawing or a cutout from a garden catalogue, laminating them and then mounting them on lollipop sticks.

and, for this purpose, untanned leathern thongs, or pieces of catgut, are preferred; silver or lead wire may also be used, the former for select plants." His wife Jane, baulking at the idea of using silver, recommended to her readers that, in the wilds of the orchard, they confine themselves to cast-iron labels.

Metal botanical labels ensured that the name of new fruit trees (or in this instance hybrid tea roses) were not lost.

that were not only insoluble in water "of any temperature," but could be marked up in ordinary ink. "Such writing, even after years of exposure, will be perfectly legible" provided no "Fancy Inks" were employed, Tebbs assured its customers.

The horticultural market was soon flooded with tin, zinc and cheap but fragile ceramic labels, along with a variety of slate and lead markers, the latter sometimes cut into thin strips that could be fastened like a cat collar around the stem of the plant.

The information carried on the label depended on the taste of the gardener. When Meriwether Lewis sent Thomas Jefferson sixty specimens of plants from North Dakota in 1805 (Jefferson was fastidious about his own labeling),

These Birmingham-made, zinc plant labels served the gardener in the days before plastic labeling.

By now botanists were adopting ever more ingenious ways of marking plants. In the Glasgow Botanical Garden one gardener devised a design for a metal spike with a small box mounted on top. The plant name was inscribed on a slip of paper and placed in the box, face up, with a glass pane puttied in place for weather protection. At the great Palazzo Pitti in Florence, meanwhile, Italian gardeners placed paper labels in special sealed glass bottles alongside their plants. The method did not impress Loudon: "A complex mode and one which can only succeed in climates like Italy."

The quest for durable and economic methods of labeling continued through the 1800s. William Cooper, "Horticultural Provider of Old Kent Road, London," was one of many Victorian suppliers who offered metal labels, ready-stamped with plant names, together with a range of helpful "botanical tablets" reminding visitors to "Keep Off the Grass" or pointing the way "To the Conservatory." A rival outfit called Tebbs marketed their "registered combination labels"

he informed the president that "these [plants] are accompanied by their respective labels expressing the days on which obtained, places where found, and also their virtue and properties when known." Gertrude Jekyll, who liked simple, triangular tabs or markers that could be popped beside the plant, preferred to keep things brief. One of her surviving labels records concisely "*Dianthus gallicus*" in curling copperplate script.

LOOKING TO THE FUTURE

Accurate labeling of stored orchard fruit was vital. Some varieties kept longer than others, and it was essential for the householder to mark their respective names in the fruit cellar, even if it was no more than a scribble of chalk on the side of the fruit box. Similar attention was paid to the contents of the ice house, the insulated storehouse packed in winter with river or lake ice where garden produce was kept in cold storage all year round. John Evelyn was one of the first to mention "conservatories of ice and snow," in 1693. William Cobbett, the bombastic and witty 19th-century author, was not a great advocate for its virtues: "A Virginian, with some poles and straw, will stick up an ice-house for ten dollars,"

he declared, adding: "I do not pledge myself for the complete success, nor for any success at all, of this mode of making ice-houses." (He did admit to never having built one himself, but believed that a redundant ice house could usefully serve "as a model for a pig bed.")

George Washington received a description of an ice house built by Robert Morris at the original presidential house in Philadelphia around 1800. The underground chamber remained at a constant temperature and, wrote Morris, "the Ice keeps until October or November and I believe if the Hole was larger so as to hold more it would keep until Christmas." Washington would later construct an ice house at Mount Vernon based on Morris's model.

Ice houses and fruit cellars fell out of fashion with the invention of the freezer by Clarence "Bob" Birdseye. Bob Birdseye had picked up the art of fast-freezing game and vegetables from First Nation Canadians while he worked as a fur trapper in Labrador. Beginning in 1916, he devised all kinds of experimental fast-freezers, including one mounted on a truck and designed to quick-freeze crops in the field as they were harvested. However, Birdseye struggled to find

18th August. 1658. Sir Ambrose Browne, at Bletchworth Castle, in that tempestuous wind. It continued the whole night, and, till three in the afternoon of the next day, in the southwest, destroyed all our winter fruit.

John Evelyn, *The Diary of John Evelyn* (1901)

WASTE NOT—WANT NOT

PREPARE FOR WINTER

Save
Perishable Foods
by
Preserving Now

Before the freezer, bottling and canning were the best way to preserve surplus fruit and vegetables from the garden.

backers for his venture until the wealthy heiress of a food-processing company sampled one of his frozen geese on board her yacht as she sailed down the Massachusetts coast. Her firm bought Bob out, changed the brand name (slightly) to Birds Eye and spent decades persuading shop-keepers and housewives to invest in the freezer.

Labeling the harvest home-stored in the freezer in clear plastic bags was less of a problem than home canning, where mistakes were only revealed when the lids were removed. Home-canning garden produce can be dated back to confectioner Nicolas Appert and his experiments with preserving food in old champagne bottles during the 1780s. Appert set up the world's first preserving factory at Massy, Paris, protecting his 40 women workers from the alarming number of exploding bottles by storing each vessel in a carefully labeled canvas bag. The canner's patron,

the restaurateur Grimod de La Reynière, boasted that, thanks to Monsieur Appert, his customers would soon be enjoying the month of May in the depths of winter. When the French government realized the potential of Appert's invention (provisioned with bottled food, their naval vessels could stay at sea longer and their *matelots* enjoy a better diet), they commissioned him to reveal all in *L'Art de conserver, pendant plusieurs années, toutes les substances animales et végétales* (*The Art of Preserving, for a Number of Years, All Animal and Vegetable Substances*). It appeared in 1810, just as Londoner Peter Durand patented the process and began preserving food by Appert's methods, but using cans rather than bottles. (Appert's business failed and the inventor died in poverty. His countrymen, however, did not forget his bottled vegetables, still a staple line in French food shops today.) Around a century later, the craft of canning (and scrupulous labeling) was brought home to U.S. gardeners as Kaiser Wilhelm II launched the Great War. Under the rallying cry "Every Garden a Munition Plant," America's National War Garden Commission offered free advice on canning and drying vegetables. "Can Vegetables and Fruit—and Can The Kaiser Too."

Thermometer

"Is the fruit cellar too hot?" "Has a
frost spoiled my apple crop?"
Temperature is the gardener's
constant concern, and the arrival of
a reasonably reliable thermometer
was eagerly anticipated.

Definition

Traditionally a glass tube filled with
mercury, used to record air, soil and
compost temperatures.

Origin

From the French *thermomètre*,
formulated in the 1620s by Jean
Leuréchon from the Greek *thermos*
meaning "hot" and *metron*, "measure."

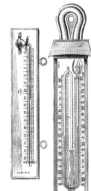

T he convoluted history of the thermometer embraces a remarkable international community of individuals from Egypt, Persia, Italy, Holland, Sweden, Denmark, Poland and Britain. Never the invention of one single person, it was always a device in development, worked on by philosophers, a polymath, a handful of engineers, mathematicians and physicists, physicians, astronomers, a physiologist, and a Grand Duke.

Philo, a Jewish philosopher based in Egypt at the height of the Roman empire, is acknowledged as the creator of the first form of thermometer. He used his equipment—a jug of water and a tube with a hollow bulb at one end—to observe the effect of warmth on the air: bubbles emerged in the water as the sun warmed the tube and the air expanded. Around the same time, the mathematician and engineer Hero of Alexandria was also studying the consequences of temperature and planning such a device for medical use. Abu Ali Ibn Sina, known in the west as Avicenna, was an 11th-century Persian polymath, regarded as among the most influential and celebrated multidisciplinarians of the Golden Age of Islam. Well versed in medicine, philosophy, theology, physics and poetry, he is credited by some sources with the invention of the air thermometer, which indicated the rise and fall of the temperature.

A standard garden thermometer, left, beside a Six maximum and minimum thermometer named after James Six.

But the invention of the air thermometer is also attributed to the Italian physician Galileo Galilei, who conceived a "thermoscope" in 1593 for registering, but not actually measuring, changes in air temperature. His contemporary, Santorio Santorii, was an Italian physiologist, professor and physician who, for 30 years, while swinging in a weighing chair beneath an immense balance, had steadfastly weighed himself, together with all his food, drink, urine and feces. His studious medical research made him famous all over Europe, but it was his obsession with describing phenomena in numerical terms that had a significant bearing upon the thermometer. It was almost certainly Santorio Santorii who was the first to bring numerical calibration to the device, in 1612.

A QUESTION OF DEGREE

The thermoscope now had a scale of measurement. Unfortunately, it was still inaccurate. Apart from his anthologies of occult philosophy, the English physician and mystic Robert Fludd succeeded in creating a thermometer with a scale in 1638, but it was still susceptible to variations in air pressure. European developments were continued by the Grand Duke of Tuscany, Ferdinando II de' Medici, who was obsessed with new technology. In 1654 he invented the very first enclosed liquid-in-a-glass device, using alcohol.

Daniel Gabriel Fahrenheit, however, a Polish-born physicist and accomplished craftsman, is also credited with the earliest thermometer to use alcohol—in 1709, during a winter of such brutality that it provided him with the measure of zero. He decided on a mercury-in-glass arrangement in 1714, the first recorded use of this element in the device, and the temperature scale he determined on his model of 1724 has caused it to be regarded as the first modern thermometer. Fahrenheit built on the earlier efforts of Ole Christensen Rømer, mathematician to the Danish court, who had devised a scale utilizing a thermometer full of red wine.

Fahrenheit's scale, which went on to become the standard in most English-speaking countries until the 1960s, is still the norm for North American gardeners. The Swedish astronomer Anders Celsius, however, recognized early in his career the need for a common scale. After careful experimentation he submitted his temperature scale to the Royal Society of Sciences in Uppsala. (He had initially given the scale the name centigrade,

The Swedish astronomer, Anders Celsius, 1701–1744.

from the Latin meaning "a hundred steps.") The year was 1742. In 1745, a year after Celsius died from tuberculosis, Carl Linnaeus, who had been busy reclassifying the living world (see p. 96), reversed Celsius's scale (originally with the boiling point at zero) and patented its use for glasshouse thermometers.

In 1782, the English scientist James Six invented a maximum–minimum thermometer that still bears his name, a device that registered

Taking an accurate reading of the soil temperature can help gardeners choose the critical time to sow their seeds.

temperatures over a given period of time. One of his mercury thermometers was almost certainly in the possession of the meticulous Joseph Taylor, author of *The Complete Weather Guide* (1816). He held up an admonishing finger and provided a graphic description of what happens when the gardener ignores a fall in temperature. "The knowledge of the exact degree of cold in the winter is of consequence to the cultivator: for it has been observed, that when the frost is so keen that the thermometer sinks fourteen degrees on Fahrenheit's scale, most succulent vegetables are thereby destroyed, such as almost all the cabbage or kale tribe, turnips, etc.; their juices being then frozen hard, their vessels are thereby torn asunder or split, so that when the thaw comes on, the whole substance, for instance, of turnips and apples, runs into a putrid mass."

Mrs. Loudon, the eminent Victorian author, in 1841 counseled that "no amateur should attempt to grow plants in a glasshouse or stove, or even in a hotbed, without being provided with a thermometer." The device, she wrote, needed to be "a very ingenious one, with a long tube for plunging into the ground . . . for ascertaining the heat of a hotbed or tan-pit." She may well have been referring to an advertisement for Pritchard's Garden Frame Thermometer in her *Gardener's Chronicle* for that year. This was "inclosed in a strong glass case, mounted with brass, so that it can be inserted into the earth, without danger. It will be found of great value for Mushroom-beds,

and the striking of delicate Flower Plants in frames." Priced at just 16 shillings, the frame thermometer came "with printed book."

The long tube inserted into the soil is still with us today, nearly 200 years later, but the ingenuity now comes in the form of handheld digital probes and wireless compost monitoring systems, linked to thermostats and humidity sensor management set-ups for day and night automatic control.

A historic film clip from 1960, filmed for British Pathé at Norwood Hall Institute of Horticultural Education in Middlesex, shows an aproned gardener entering a glasshouse filled with exotic, subtropical plants. On the wall close by is the U-tube of a maximum–minimum thermometer. Norwood Hall was part of a post-World War II project to combat food shortages, and the thermometer was critical in helping the gardeners maintain the right temperature. Critical and proven: in the year the film was made, the Institute produced some 250 bananas.

Scarecrow

The disciplines of pest control and sculpture form an unusual alliance in the scarecrow. While the outstretched arms of the straw-stuffed bogeyman, draped in an old jacket and wearing grandma's bonnet, provide a handy perch for the pigeon, no fruit yard is complete without him. Or her.

Definition

The scarecrow or *Vogelscheuche* is designed to keep birds at bay in the garden.

Origin

The scarecrow took the place of the women or children once paid to frighten off birds.

I t is the old story: "That which the palmerworm hath left hath the locust eaten; and that which the locust hath left hath the cankerworm eaten; and that which the canker-worm hath left hath the caterpillar eaten." The battle to keep hold of the fruits of the vegetable garden or orchard has been raging since biblical times, as this line from the Book of Joel indicates.

The avian world is a constant problem throughout the garden and the German *Vogelscheuche*, the Danish *fugleskræmsel*, the French *épouvantail*, or that defender of the Scottish potato crop, the *tattie bogle*, are all designed to frighten the birds away. Scarecrows, especially if they can be persuaded to flap about in the breeze, are sinister figures. The Japanese gave their *kakashi* an extra dimension: a sour smell. Bamboo poles were draped with old rags and bits of offal, which, when set on fire, caused a foul odor to drift across the paddyfields.

As far as the Roman writer Columella was concerned, the solution lay in placatory sacrifices including the "blood and entrails of a sucking whelp," the skinned head of an Arcadian ass and "night-flying birds on crosses." His fellow Romans did, however, make use of mannequins to scare away birds. These were naked figures of the god Priapus, who carried a scythe in one hand, the hopeful token of a good harvest. Centuries on, their descendants were still scaring pesky Mediterranean birds with bird-scaring poles topped with sun-bleached animal skulls.

Engraving called Guarding the Corn Fields *where a Native American woman beat pots with sticks to scare the birds away from fields of corn.*

One of the scarecrow's more sinister manifestations is the character of Dr. Syn, the fictional creation of writer Russell Thorndike, brother of English actress Sybil Thorndike. Thorndike's 1915 novel *Doctor Syn* (remade by Walt Disney as *The Scarecrow of Romney Marsh* in 1963) featured the macabre figure of the scholar turned smuggler, Christopher Syn, who roamed the lonely marshes of southeast England disguised as a scarecrow.

Using a scarecrow to scare away birds had disadvantages: it allowed the locusts, cankerworms and caterpillars to multiply along with the gardener's arch-enemies, slugs and snails. As the naturalist Gilbert White wrote, somewhat bitterly, in his diary during 1759: "Sowed a pint more of dwarf kidney beans in the room of those that were devoured by snails." There were plenty of ingenious ideas for dealing with this particular menace: one Arab gardener advocated sprinkling ashes "from the public baths" around sensitive plants. Columella suggested extracting the sap of houseleek or horehound and drenching seeds and plants with it, while the 16th-century English author Leonard Mascall advised popping the pests into a poisonous bath of ashes and unslaked linden. Alternative infusions included walnut leaves, ground-up shells, crushed tobacco leaves and vinegar or linden water, which, one writer promised, was guaranteed to "annoy them." Anthony Huxley in his *Illustrated History of Gardening* discovered an 18th-century Londoner who kept four pet seagulls to deal with "little beasts injurious to kitchen gardens."

BEERY DEATH

Other defenses included encircling plants with a zinc or copper collar, sprinkling salt on slugs, creating a defensive wall of prickly barley-straw heads or simply gathering the creatures up and throwing them over the garden wall (although amateur experiments that involve marking the snail's shell with nail polish suggest that the mollusc possesses remarkable homing instincts). The organic gardener

A host of treatments have been devised to defeat the gardener's traditional enemies, slugs and snails.

Lawrence Hills recommended a beery slug trap. "The traditional soup plate, wide and shallow, sunk level with the ground and filled with a mixture of 1 part of beer to 2 of water, sweetened with 1 dessertspoonful of Barbados sugar to 1 part of the mixture."

The chemical solution was metaldehyde slug bait, formed into pellets and scattered around susceptible plants, but since the World Health Organization has classed it as a "moderately hazardous" pesticide capable of causing secondary poisoning to wildlife and pets, many gardeners regard it with suspicion.

Slugs and snails are not the only enemies in the garden, and horticulturists have devised a host of traps, cages and barricades to deal with animals ranging from rats and mice to deer and elephants. While the scent of chili pepper is said to be effective against elephants, electric fences and repellent sprays have been employed (with limited success) to deal with wallaby troubles in Australia's Northern Territory, wild boar rooting up lawns in England's Forest of Dean, or moose munching lilac buds in Alaska. "Nothing stops a moose. Nothing!" wailed one New Hampshire correspondent.

The worst offenders are species introduced inadvertently or deliberately into a new habitat: the rabbit in Australia (brought in with the First Fleet as a source of food in 1788), the Australian possum in New Zealand (introduced as a fur-farm animal in 1837) and the otherwise

Ten Ways to Encourage Birds

+ Hang up a bird feeder.

+ Set a birdbath out of reach of cats.

+ Keep a small garden pond.

+ Mount an insect hotel—an open-fronted box packed with small pieces of wood— on a pole.

+ Create a little log pile to encourage insect populations.

+ Establish a simple bog garden with a shallow tray of damp moss and soil.

+ Plant bee- and butterfly-friendly plants.

+ Leave a corner of the grass unmown to encourage both insects and wildflowers.

+ Plant fruit trees for perching and food.

+ Introduce a scarecrow—even if it simply provides a perching point for the birds.

slugs, grubs, and other insects [*sic*], as soon as they appear." He echoed the misguided view of an unidentified writer from the Middle Ages who insisted "wormes that . . . waste my herbes, I dash them to death."

Loudon also advised that "a few cats domiciled in the back sheds of hothouses will generally keep a walled garden free of mice." This arrangement had its own drawbacks, not least "the Cats scattering their Ordure all about, and then scraping the Earth to cover it," as Henrik van Oosten explained in *The Dutch Gardener; or, the Compleat Florist* (1703).

OVERKILL

In the end, the gardener's war on the natural world has had an adverse effect. Plunging populations of wildlife species, including not only birds, but also bees, butterflies and other pollinating insects, has led to one of the most significant changes in attitudes between contemporary gardeners and those of the 19th and early 20th centuries. While gardeners in the past used every means to rid the garden of wildlife, their descendants began making every effort to conserve it, providing frog-friendly ponds, insect log shelters, bee "hotels" and bird feeders. The scarecrow has, after all, become the most non-invasive and environmentally friendly way to protect garden crops.

lovable hedgehogs foolishly released on the formerly hedgehog-free Outer Hebrides during the 1970s to control garden pests. (The spiny animals ended up eating the eggs of ground-nesting birds.)

The hedgehog's fondness for eating earthworms, however, endeared it to John Claudius Loudon. The garden writer regarded worms as a menace equal to slugs and snails, and advised the gardener to "gather by hand all worms, snails,

Structures & Accessories

The imagination knows no bounds when it comes to filling the garden space. And yet, from "Hitler's" cloches to grand glasshouses and from humble potting sheds to pretty, terracotta planters, garden paraphernalia usually has an underlying practical purpose.

Potting Shed

Home to an assortment of garden tools, crop plans and seed catalogs, the potting shed is both a wet-weather retreat and a place for rest and recreation. Largely a product of the 19th-century gardening boom, the garden shed is also the second most dangerous place in the home.

Definition

A small building for storage and maintenance of garden equipment.

Origin

"Potting shed" from Old English *pott*, a cylindrical vessel, and *shadde*, first used in print by William Caxton in 1481.

174

T he potting shed is a place of preparation, reference, reflection, respite and shelter, the place where things begin and have their continuity. It embraces creativity of all kinds, from the ardent back-garden hobbyist's to that of the professional artist. The shed has been host to parallel worlds in which things flourish. The birth of the wind-up radio in the mind of inventor Trevor Baylis took place in a shed, while Mark Twain, Virginia Woolf, Agatha Christie, Arthur Miller, Dylan Thomas, Jeanette Winterson and Louis de Bernières have all worked in their sheds. The Irish playwright George Bernard Shaw and garden-lover and sculptor Barbara Hepworth both installed couches in their sheds, as did the 19th-century gardeners at Tyntesfield House in Somerset; they slept in bunks positioned above their workbenches.

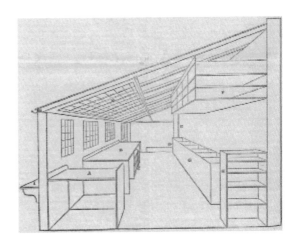

Mr. Beck's potting shed was a triumph of form and function according to the Gardener's Chronicle *of 1843.*

REMARKABLE NEATNESS

The November 1843 edition of the *Gardener's Chronicle* offered an inside view of "a most convenient potting-shed belonging to Mr. Beck, of Isleworth." It was said to be "remarkable for its neatness and general arrangement, and contains within itself every convenience which the Amateur can desire." Among the practical facilities on offer were substantial benches for potting and display, "a large water tub, running on castors, which may be pushed under the bench, out of the way," Stephenson's conical boiler, a fruit loft, containers for peat, sand, crocks, and a cupboard for paint pots and tools. With the emphasis on preparedness and order, a row of bins is also mentioned, receptacles that enabled the gardener "to keep all his composts and potting materials distinct, and always ready, so that there is no time lost in hunting after this or that, and the place never needs to be unneat."

It was into such environments that, during the 19th century, teams of gardeners and boy apprentices, or individuals working alone, would engage in all the seasonal tasks required to service gardens large and small. From seed sowing, propagation, pricking out and trans- planting, to taking cuttings, dividing perennials, mixing composts or repairing tools, this was, and continues to be, the heart of the garden operation.

Helena Rutherfurd Ely, once described as the American Gertrude Jekyll, set out the practical arrangements for the tool room in her popular

The history of the garden shed may be traced back to Roman times and Pliny the Elder.

book *A Woman's Hardy Garden* in 1903. Ely, who created a five-acre (2 ha) garden at her 350-acre (142 ha) Meadowburn Farm in New Jersey while writing her popular gardening books, imposed a sense of order on her own garden tools. "On a particular shelf in my tool-room I keep my private trowel and flower scissors, to which are attached long red ribbons as a warning of "Hands off!" to others."

She had a kindred spirit in the Australian garden designer Edna Walling, who drew up many garden plans for clients. However grand the setting, Walling invariably included a potting shed somewhere on the grounds. She was also, like Ely, wary of lending out her own garden tools after a friend borrowed a treasured fork. When it was returned with a prong missing, "I silently vowed: 'Never again.'"

It is difficult to pinpoint the origins of the potting shed. The Roman author Pliny the Elder, who confessed that much of his writing took place in the context of a garden, may well have

retired to a garden retreat to record his thoughts. During the reconstruction of the site of Fishbourne Roman Palace in the south of England—a luxury villa surrounded by one of the earliest gardens ever found in Britain—a Roman potting shed was also recreated. Designed to service the formal arrangement of bedding trenches, shrubs and fruit trees as well as an assiduously cultivated vegetable garden, the shed was furnished with bench and shelving, baskets and ceramic pots and a variety of Roman gardening tools.

TOOLS IN ACTION
Shed Organization

When siting a new shed in the garden, select a level area and choose a building that is large enough to fill the space. If you're installing electricity, follow local codes for wiring, circuit breakers and the like. Having a water supply and sink inside the shed is a very good idea, as is a an impermeable floor covering such as vinyl. Keep seeds secure in an old but working fridge. After cleaning, plunge tools into a bucket containing a blend of oil and sand, to help prevent rust. Build in time for that annual spring cleaning, after the clutter and disorder that can accumulate during the growing season.

In large Victorian walled gardens the number of outhouses and sheds could run into double figures, with garden offices, seed stores, a bothy for the water boy, tool stores, stables and root cellars along with the potting shed. One such building was reconstructed at the Victorian Productive Gardens at Heligan in Cornwall. The potting shed in Heligan's pre-World War I Melon Yard was equipped with long potting benches, lit by the light of the window, under which the various composts were stored. On the opposite wall, rows of pegs were positioned at different heights for the gardeners' rakes, hoes and other tools beside the mighty rack built to store rows of clay pots, stacked on their sides according to size.

The potting shed was also home to the garden catalog, which from the mid-1800s marketed a range of potting sheds, follies, glasshouses and rustic summer houses for the family villa gardener. The gardener, for example, could spend a profitable Saturday afternoon in the 1860s assembling his Portable Hexagon Summer-House (£22), "easily put together by two persons in one hour." If, like most people, he was merely renting the property, he could opt for one of William Cooper's "Tenant's Fixtures," a shed that was made in sections and could "easily be put together or taken down and removed."

The craft of the shed, or its smarter sister the summerhouse, reached its apotheosis on the allotment. Some served as home away from home for a husband or wife trapped in an

The fully functioning garden shed where the average UK gardener spends five months of their life.

unhappy marriage; others even became a place of births and deaths. David Crouch and Colin Ward in *The Allotment* (1988) recorded how Nottingham's oldest allotment site at Hunger Hills was dotted with brick-built summerhouses where, in the home-hungry days of the early 1900s, many people lived. One woman had been born there and raised a family of eight.

From the 1920s onwards the garden shed has developed gradually into its present-day form, into something that is more of an intimate, personalized space. Britain hosts an annual Shed of the Year competition, and it has been calculated that a solid 5 months of the lifetime of an average UK citizen is spent in a garden shed. It is also the environment where nearly 20% of all men meet with some form of accident.

Glasshouse

For centuries the glasshouse was the preserve of the rich and infamous. Radical technological advances in the 19th and 20th centuries finally turned it into an everyday gardening tool and the scene of one of the finest sights in horticulture: a glasshouse full of cucumbers.

Definition

In its traditional form, a building with a roof and sides made of glass set in frames, for growing plants in a protected environment.

Origin

The term is first found in use in the 1660s.

T he unpopular Roman emperor Tiberius alledgedly possessed a voracious appetite not only for debauchery, but also for cucumbers. The "cucumber" referred to by Pliny the Elder and Columella was probably the snake melon, a sweet fruit highly regarded in Rome at the time and known today as the Armenian cucumber or *faqous* (*Cucumis melo* var. *flexuosus*).

Sick and enfeebled, Tiberius was instructed by his team of physicians to eat a cucumber every day of the year. To guarantee supplies, he had a special *specularium*, a house for growing plants, roofed with oiled cloth, constructed around AD 30. An alternative to oiled cloth was mica, a silicate of the mineral gypsum, which possessed a crystalline structure that allowed it to be split into thin, translucent slices. Pliny described plants being "placed under the protection of frames glazed with mirror stone," one of the vernacular names for the mineral. In the winter months, fires were kept continually burning outside of the stone walls to warm the house, while manure hotbeds (*see* p. 89) were also used to maintain a high temperature.

GIARDINI BOTANICI

Roman gardeners continued to improve the design of their specularia and filled them with crops of roses and grapes, although it was not until the 13th century that the Italians devised the first glasshouses that we would recognize today. These were the *giardini botanici* or

TOOLS IN ACTION

Glasshouse Practicalities

The roof covering of a glasshouse determines how much light will enter the space. Glass is the traditional material, and better at heat retention than plastic sheeting. Compared to clear glass, frosted panes permit a more even distribution of light. There are, however, many plastic alternatives, which are often shatter-resistant and have the advantage of being cooler than a traditional glasshouse during the summer months. Arched plastic greenhouses have become a popular and practical option. The plastic film covering is cheaper than glass, lasts from three to five years or longer, and is available with a coating that reduces dripping.

botanical gardens (the Vatican built one of the first), built to house the new plant species being brought back to the Mediterranean countries as their empire builders explored new lands. The glasshouse technology, together with the plants, spread initially to Holland and later to England and France in particular. Keeping the glasshouse warm, however, remained a problem.

By the 15th century, the glassmakers of Murano near Venice were producing a transparent compound that found its place in what were

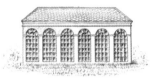

A 17th century English glasshouse, right, alongside what was described as America's first glasshouse, erected in New York in 1764.

now to be known as glasshouses or conservatories (the latter term is used for a building where the plants are grown in beds or borders rather than in containers).

In 1599 Jules Charles, a French botanist, designed and built a glasshouse in Leiden, Holland, and used it to raise tropical plants for medicinal purposes, along with citrus fruits. The fondness and fashion for these tropical fruits led to the creation of orangeries and drove the development of ever more ingenious designs. In 1619 the French Huguenot Salomon de Caus, an enterprising engineer who worked on the garden plans for Heidelberg Castle in Germany, was assembling removable wooden walls in a glasshouse for exotic plants.

During the 17th century, glasshouses, many of which now possessed heating flues for overwintering, were used in the summer for entertaining, since the majority of the plants had been moved outside. The glasshouse had become a status symbol.

The development of the glasshouse in the United States was inextricably linked to the slave trade. Andrew Faneuil, referred to in contemporary accounts as a merchant, was a more-than-financially comfortable Boston entrepreneur whose fortune had been largely built on the sale of Africans to the West Indies. He was acknowledged as the owner of the first American glasshouse, in which he grew fruit, in 1737. The building has long since disappeared and the only

18th-century American glasshouse still standing is the Wye Orangery, from 1785, part of a plantation on the Eastern Shore of Maryland. It was here that Frederick Douglass, the African-American statesman, spent some of his childhood in slavery. In some instances slaves not only managed the glasshouses, but had their living quarters within the buildings as well.

In Britain the glasshouse was gaining momentum. When a particularly unpopular tax on glass was finally lifted in 1845 it contributed to a boom in both glassmaking and the manufacture of glasshouses. To let in more light, narrow wrought-iron bars were now available that could be used in place of a wooden framework to support the glazing. And, although the structures were relatively lightweight, they could be cast, in what was called the "curvilinear principle," into vast, arched spans. In 1840, at the stately home of the Duke of Devonshire, Chatsworth House in Derbyshire, head gardener Joseph Paxton used this principle (though working in wood) to assemble a glasshouse that was christened, with typical Victorian hyperbole, the Grand Conservatory. Grand it was, occupying an acre (0.5 ha) of ground and heated by enough hot water piping to have carried a water supply to neighboring Bakewell and back (6 mi./9.5 km). The entrance was so wide that Queen Victoria, when she visited in 1843, could be driven in her carriage straight into the building.

Queen Victoria's royal patronage also extended to the Palm House at Kew's Royal Botanical Gardens, among the first of many glasshouses to function as a public space. Built between 1844 and 1848 and borrowing on ideas from the shipbuilding industry, the Palm House rested on a frame of sixty cast-iron ribs, obviating the need for intrusive supporting pillars. Larger still, one of the most impressive glasshouses of any period was initially sketched out on a sheet of blotting paper by Joseph Paxton. The magazine *Punch* dubbed it the Crystal Palace, and it was built for London's Great Exhibition of 1851.

Three times longer than St. Paul's Cathedral and fitted with some 900,000 sq. ft. (83, 613 m²) of glass, it was so vast that it could house some of Hyde Park's tallest elms.

This was the era of the giant conservatory, of the Palm House at the Berlin Botanic Gardens, of the Royal Greenhouses of Laeken built for King Leopold II of Belgium, and of a 230-yard (210 m) vine house at Buffalo, New York, housing over 200 vines. These great airy structures echoed the glazed halls of the new railway stations, and they were accompanied by the more intimate, domestic form of the glasshouse room. "The marvel of this wondrous palace," enthused a correspondent from *L'Illustration* in 1876, reporting on the Parisian mansion of a financier, "was indisputably the 'winter garden'. Our great ladies of fashion found refuge there to avoid the crowds. Only a king or a banker would surely dare to surround himself with such sumptuousness."

By 1905, Chatsworth in the English Midlands possessed almost 2 acres (0.8 ha) of glasshouses. However, in the postwar years practicality, cost-effectiveness and affordability became the hallmarks of the glasshouse. By the 21st century, building materials such as fiber-glass, acrylic, polycarbonate, aluminium, polyethylene and PVC meant that glasshouses were no longer the preserve of the elite. Now any homeowner could enjoy what James Shirley Hibberd described as "a houseful of cucumbers . . . one of the finest sights in the whole range of horticultural exhibitions."

The biggest glasshouse of its time, Crystal Palace in London was three times longer than St. Paul's Cathedral.

Cloche

The old-fashioned iron-and-glass handlight (or hand glass) cloche has enjoyed a revival as a decorative ornament. The glass cloche was a practical tool for protecting plants, but material changes, from glass to plastic, brought real benefits to the garden.

Definition

A glazed cover placed over a plant to provide a warm environment. The single cloche and the barn cloche can extend the growing season by a week or two.

Origin

From the French word for a bell, the cloche was originally made from hand-blown glass.

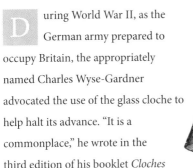

Bell cloches were claimed to be among Britain's secret weapons during World War II.

During World War II, as the German army prepared to occupy Britain, the appropriately named Charles Wyse-Gardner advocated the use of the glass cloche to help halt its advance. "It is a commonplace," he wrote in the third edition of his booklet *Cloches Versus Hitler* (1941), "that the battle of the Atlantic is being fought not only on the sea and in the air, but also on land in every garden and allotment in the country."

Liberty ships were needed to ship troops and armaments rather than food to Britain from America, and it was the duty of every citizen to grow as many vegetables as they could. "Cloches are not luxuries, but rank with seeds, manures and garden tools, as part of the equipment essential for successful all-the-year-round vegetable growing."

Young plants at the start of the growing season, and plants still ripening at the end of it, are vulnerable to the cold. A cloche or a cloche tunnel prolongs the growing period, creating a modestly warmer microclimate for the plants. Modern cloches, which also protect crops from pests and snow damage, range from replicas of 19th-century designs such as the bell, the handlight and the lantern frame (square or octagonal constructions with lids that can be lifted off in the heat of the day) to plastic-covered wire-framed types and even recycled plastic beer bottles with the base cut away. Tunnel cloches vary from a tent-like construction of glass sheets in wire frames to wire hoops covered with clear plastic sheeting, and semi-rigid plastic frames that are light enough to be carried around the vegetable garden (and blown about on stormy nights).

As well as extending the growing season, cloches will help to harden off plants (acclimatize them to cooler temperatures), dry crops such as onions, garlic and potatoes after lifting and before storing, and blanch crops such as endives, celery and rhubarb. Blanching (keeping the crop growing in semi-darkness to slow photosynthesis and produce a pale flesh) can be achieved by covering the crop with blind cloches (which may be made of black plastic or of glass painted to make them opaque) or blanching pots.

The business of blanching rhubarb, the earliest fruiting crop in north European gardens, had developed into a regional speciality in the West Yorkshire "rhubarb triangle," where crops were brought on for the London market, initially under blanching pots, and later in darkened sheds heated by coal stoves. Bell cloches were marketed in some garden catalogs as being of "correct greenish glass, which does not allow the sun to burn the lettuces or cause them to run quickly to seed."

The glass of the traditional bell cloche was said to protect the plants from sunburn.

The most common material for the modern cloche, plastic, owed its origins to New York immigrant Leo Baekeland. In the early 1900s Baekeland was researching polymers (from the Greek *polus*, "many," and *meros*, "parts"), chemical compounds with a molecular structure like a chain of large molecules, each of which was formed from smaller chains. Baekeland was searching for a synthetic substitute for shellac, which is made from the secretion of the Southeast Asian lac beetle. He came up with a hard, black plastic (from the Greek *plastikos*, "able to be shaped or moulded"), which he called Bakelite. (At the turn of the 20th century, Scottish engineer James Swinburne went to patent his own version, Damard (as in "damn hard"), only to discover that Baekeland had patented his the day before.) In 1924 *Time* magazine was predicting that "in a few years [Bakelite] will be embodied in every mechanical facility of modern civilization. From the time that a man brushes his teeth in the morning with a bakelite-handled toothbrush, until the moment he falls back upon his bakelite bed in the evening,

all that he touches, sees, uses, will be made of this material of a thousand uses."

Bakelite made a modest impact on garden implements, particularly for tool handles, but by the time the father of modern plastics died (in a New York sanatorium in 1944, having become an obsessive recluse living on nothing but canned food) the war had accelerated research into

TOOLS IN ACTION

Using a Cloche

Different seeds germinate at different temperatures, but most of those in the vegetable bed require a warmer soil than many an impatient gardener realizes. Carrot seeds, for example, need a temperature of around 8°C (46°F); green beans need it warmer still, around 12°C (54°F). Cloches should, therefore, be set out at least two weeks before sowing in order to warm the soil. This has the added benefits of drying the soil (seeds can rot in wet soils) and bringing on weed germination. Weeds can then be hoed off before the seed is sown. Where damage from birds and rabbits is a problem, the cloche tunnel can be used to preserve plants at the vulnerable seedling stage. The cloche also protects plants from snow damage.

thermoplastics. They were to transform the garden cloche.

Unlike Bakelite, thermoplastics softened under heat, so they could easily be moulded. Polyurethane, polystyrene and polymethyl methacrylate (commonly known by the trade names Plexiglass and Lucite), were soon being heralded as the new miracle materials. Of particular interest to the gardener were plastic cloches and polyethylene tunnels, polypropylene water pipes, polyester water tanks, polyvinyl chloride (PVC) glasshouses and polystyrene packaging and insulation. There were even plastic flowers and plastic grass. Used PVC doors and windows, meanwhile, were pressed into service as ramshackle but serviceable sheds, glasshouses and cold frames by smallholders and home gardeners with an inclination to make the most of recycled materials.

Close to nature as they are, gardeners tend to be environmentally aware, and many feel justifiably virtuous in growing their own to reduce their food's carbon footprint, or choosing a hand tool rather than its fossil-fuel-driven equivalent. But finding a suitable, sustainable substitute for horticultural plastics, from the cloche to the plastic plant pot, has proven to be a problem: plastics spend a short time in the garden and a dangerously long time drifting around the world's oceans in a soup of plastic waste. The plastics problem has led to repeated calls for the garden industry to recycle its waste, and stimulated research into alternative materials.

During World War II, however, the only material available to followers of Charles Wyse-Gardner was glass. His *Cloches Versus Hitler* campaign was sponsored by one of the nation's main cloche makers, Chase. "In wartime, when results are vital, more than 100,000 gardeners . . . are using the Chase Continuous Cloche," declared Chase, promising that their product was not only portable, but "blast proof." And it was Chase that could claim credit for one of the most abiding slogans of the war: Dig for Victory.

This 17th century vegetable garden provided early and late crops through the intensive use of cloches, hotbeds and a protective brick wall

Terrarium

To the Victorian gardener the fern-filled Terrarium—known then as a "Wardian case"—was a charming decorative item. But the glass, brass and wood invention of Dr. Nathaniel Ward was transforming, bringing in a host of new plants that had never before survived a long sea voyage.

Definition
> An airtight case that provided a self-sustaining environment for transporting plants.

Origin
> Developed in the 1820s by a doctor working in London's East End.

T he early riser who wanders through her garden, nursing a mug of coffee and taking stock of her horticultural handiwork, cannot fail to be impressed by the number of "foreign exotics" growing there.

The traffic in plants that brought Australian eucalyptus to New England, European heathers to New South Wales or Turkish tulips to Newfoundland started in earnest in the mid-1500s, with new plants being imported into Europe from the Middle East. From 1620 the flow into Europe switched to North America, especially Virginia and Canada, with many of the introductions coming through celebrated plant hunters such as John Tradescant (1570–1638) and his son, also John (1608–1662).

The plant trade could be a lucrative business. Fuchsias were first discovered growing in the Caribbean in the 1690s, and when, a century later, an eagle-eyed plant dealer spotted a fuchsia in a sea captain's London garden, he was prepared to pay £80 for it. (The annual wage of a ship's bosun then was around £20.) The nurseryman was soon marketing rooted fuchsia cuttings for up to £20 each.

As was almost the case with the fuchsia, a plant's discovery in some distant land did not necessarily herald its introduction to the European or American garden. One of the most dramatic discoveries of new plants occurred when the Royal Navy barque *Endeavor* sailed from Plymouth, England, for Australia and New Zealand under Captain James Cook. In April, 1770, the ship landed near what would become the Sydney suburb of Kurnell, and in a feverish eight-day scramble Joseph Banks and Daniel Solander, the two botanists on board, collected an estimated 10,000 specimens, including almost 1,500 plants then unknown to horticultural science.

BOTANY BAY

The *Endeavor* struggled home, reaching England almost a year later as Banks and Solander recorded the finds in their herbaria. Cook later renamed his landing place Botany Bay as a tribute to the records kept by Banks and Solander. One, held at the National Herbarium of Victoria at the Royal Botanical Gardens in Melbourne, reads: "*Banksia serrata* L.f. Col: Banks & Solander, 1770. Loc: NEW SOUTH WALES, Botany Bay," a reference to the saw banksia or red honeysuckle.

Young fuchsia plants were changing hands for exorbitant sums in London in the late 18th century.

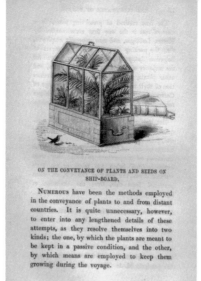

ON THE CONVEYANCE OF PLANTS AND SEEDS ON SHIP-BOARD.

NUMEROUS have been the methods employed in the conveyance of plants to and from distant countries. It is quite unnecessary, however, to enter into any lengthened details of these attempts, as they resolve themselves into two kinds; the one, by which the plants are meant to be kept in a passive condition, and the other, by which means are employed to keep them growing during the voyage.

Plants could survive for months and even years in the microclimate provided by one of Dr. Ward's glass and brass cases.

The passage of live plants on board ships like the *Endeavor*, however, was a precarious one. Salt air, seawater and a shortage of fresh water all contributed to significant attrition rates. This was soon to change.

In the 1820s a London doctor, Nathaniel Ward, set out for an early-morning walk before work. Deeply interested in natural history, Ward would scour the hills and hedgerows for interesting items, and on this particular morning he came across a chrysalis, which he popped into a glass collecting bottle along with a pinch of the earth on which it lay. Fixing the airtight stopper in place, he stood the specimen on his windowsill to observe developments. Some days later he noticed that some seedlings in the trapped soil had sprouted and grown, despite the absence of any fresh air. As long as the bottle remained sealed, the plants continued to thrive because evaporation and condensation kept the total amount of water in the bottle constant.

Appreciating the significance of his discovery, Ward commissioned some larger specimen cases, in wood, glass and brass, from a well-known Hackney nurseryman, George Loddiges. Loddiges's family had traded in exotic imports since the 1770s, especially from Australia—they were responsible for the introduction to Britain

of rhubarb and the common mauve rhododendron (*Rhododendron ponticum*). Loddiges' cases were planted up and, once again, the plants thrived.

Ward then shipped two fern-filled cases out to Sydney and Botany Bay, where they were emptied, refilled with native Australian species, and sent home. The plants survived both journeys.

Ward's work sent a whisper of excitement through London's horticultural circles. John Claudius Loudon was persuaded to pay the doctor a visit and subsequently reported in the *Gardeners' Magazine* for March 1834: "The success attending Mr. Ward's experiments opens up extensive views as to the application in transporting plants from one country to another; in preserving plants in rooms or in towns; and in forming miniature gardens or conservatories . . . as substitutes for bad views, or for no views at all."

The Wardian case, later to be known more generally as the terrarium, did indeed trigger a fashion for drawing-room ferneries and other "miniature garden" collections. One lady wrote to Ward from Bristol: "I think you must have much satisfaction in thinking how much pleasure you have been enabled to give in the world, and how often the sorrowful have been cheered by watching the fresh green vegetation near them, when illness or their occupations in life confine them to the dark smoky streets of a large town." The terrarium, however, was having a far more hard-nosed effect on the world of horticulture and commerce. The activities of some 19th-century plant hunters added a dash of excitement to the otherwise genteel goings-on in gardening

TOOLS IN ACTION
Moving Plants

Deciduous plants are best moved during their dormant period. Anything that helps reduce transpiration during the move—light pruning, tying the branches loosely together, keeping the root ball intact—will help. Larger plants and small trees can be prepared for the move during the previous growing season by trenching around the plant and filling the trench with sharp sand. This allows the plant to make fibrous feeding roots. Ideal conditions for moving are on a windless, dull day. Small plants and cuttings can be protected during a move by replicating the terrarium: place them in a plastic bag or on a seed tray covered with a clear propagator lid. Many perennial plants are more easily lifted and divided.

circles. David Douglas, responsible for the introduction of the lupin, penstemon (beard-tongue) and Douglas fir, died in 1834 before the arrival of the Wardian case. Seven years later Joseph Hooker was able to make the most of the cases when he shipped numerous plant specimens back from a trip to New Zealand. Another nurseryman who benefited was Robert Fortune. In 1843 he was sent to China by that august body, the Horticultural Society (founded in 1804 for "the improvement of horticulture") armed with terrariums, pistols and instructions to bring back everything he could. He arrived home with tales of daring exploits and with future garden favorites such as *Weigela rosea, Lonicera fragrantissima* (winter-flowering honeysuckle) and *Mahonia japonica*. Fortune later smuggled 20,000 tea plants in terrariums out of China to what would become the great tea plantations of Assam in northeastern India. (The invention came too late for the valiant French naval officer Gabriel de Clieu who, back in 1720, had managed to ship a single coffee tree from France to the island of Martinique, maddening his fellow passengers by sharing his meager water rations with the tree when their ship was becalmed.)

When Ward published his findings in *On the Growth of Plants in Closely-glazed Cases* (1842), he could justifiably boast: "There is not a civilized spot upon the earth's surface which has not, more or less, benefited by their introduction."

Plant Container

A raft of contemporary books and
programmes has been devoted to
the topic of container gardening.
Yet, from the Roman window box to
the French *jardinière*, this is one of the
most venerable crafts, practiced with
due diligence across the world from
Egypt to ancient China.

Definition

A receptacle or enclosed vessel in
which a plant or tree is grown.

Origin

The growing of plants in containers
dates back more than 35 centuries; the
word "container" is of Latin origin.

F inely featured, charismatic and elegant, Hatshepsut was one of the very few women to hold power in ancient Egypt, a queen who nevertheless wore all the regalia required of a male pharaoh—including the false beard. Thirty-five centuries ago, in 1500 B.C., her ambassadors returned from a voyage to the fabled land of Punt with gold, ivory, incense, ebony and monkeys, and perhaps more significantly, 31 myrrh trees, living trees whose roots were mindfully packed in baskets to withstand the journey. This amounted to the first known instance of containers being used to transport trees.

Hatshepsut gave orders for her trees to be planted in the courtyards of her own mortuary temple on the banks of the Nile. This stunning architectural setting honored the queen and raised container gardening to a regal level. Archaeologists have exposed tree pits in the lower court area, and it is likely that the terraces were adorned with sandstone troughs filled with frankincense—perfumes and essences were important aspects of Egyptian culture for both the living and the dead.

Roman gardeners also mastered the craft of container gardening, although on a more modest scale. The garden, a central feature of the Roman family villa, was gracefully laid out in the *perystylium*. This was a garden courtyard bordered by a covered and paved pathway, patterned with a set of evenly spaced columns and reminiscent of some modern Mediterranean

Trees, their roots bound ready for transporting, are prepared for Queen Hatshepsut's mortuary temple on the Nile.

homes. Planters could be used as dividers between one house and its neighbor: at a house discovered in Pompeii, the seaside town near Naples buried under volcanic ash in A.D. 79, flower boxes had been filled with earth and planted with blooms and creepers to create a curved dividing wall. Even in homes where there was no space for a peristyle, the housewife would emulate her wealthier neighbors, setting out plants such as acacias, acanthus, anemones and perhaps a little apricot tree in terracotta pots around the house. If there was no other room for plant pots, there was always the open window. Pliny the Elder commented in his *Natural History* on how the poor used window boxes to conjure up the rural landscape: "In former times the lower classes of Rome, with their mimic gardens in their windows, day after day presented the reflex of the country to the eye."

Flowerpots were popular too during the medieval period in Europe, especially for the carnation grower. Carnations were grown on in special ceramic pots or narrow-topped jugs, crocks or urns, as shown in many of the paintings of the time. In Carpaccio's painting of *The Dream of Saint Ursula* (1495), for example, two potted plants stand conspicuously in the window space: the artist has portrayed the householder's efforts to build a framework around the top of one pot to support the carnations growing inside.

TRAY SCENERY

Japanese paintings, too, reveal the use of containers, especially in the art of *bonsai*, growing miniature trees in pots or trays. An early representation of dwarf trees in pots dates from A.D. 1195, while a later scroll of 1309 shows miniature landscapes in dishes and wooden trays, displayed upon contemporary-looking benches. The concept came from China in the 6th century A.D., where gardeners had long relied on container-grown plants to link the house with the natural world outside. Plants were positioned around the edges of open courts in order to alleviate and soften the severe profiles of the rocky outcrops behind.

China's gardeners were also in the habit of using the stone supports of pagodas to display

A servant, pictured in a 1300-year-old mural from a Tang Dynasty tomb bears a miniature landscape in a ceramic dish.

plant-pot stands together with small trees and shrubs grown in low, ceramic containers like shallow trays. These were examples of the Chinese art of *penjing* or "tray scenery," and it was these little landscapes with their miniature rocks and trees that were taken to Japan and that influenced the growth of bonsai. From the mid-1400s until about 1800 Japan imported Chinese-made bonsai trays in unglazed, purple-brown ceramics. The poor man's alternative was to plant the tree in an abalone shell, the spiral mother-of-pearl casing of an edible sea snail.

A bonsai forest grown in trays at the gardens of the pagoda Yunyan Ta on Tiger Hill at Suzhou in China.

In 16th-century Suzhou, China's version of Venice, imperial envoy Wang Xiancheng retired from his turbulent life as a magistrate to celebrate the art of container gardening at what became known as The Humble Administrator's Garden. Here large, glazed ceramic urns functioned as ponds for growing fragrant aquatic lotus flowers. Later, in the 1850s, the ceramic jar makers who produced similar types of vessels moved from Swatow (Shantou) in south China to Malaysia, establishing a trade in planters that continues to this day.

The British, by the 16th century, were becoming increasingly passionate about their gardens, and Thomas Hill counseled the readers of his *Gardener's Labyrinth* on how to recycle their old pots. To find containers in which to sow cucumbers, he suggested that they "fill up old worne baskets and earthen pans without bottomes, with fine sifted earth tempered afore with fat dung, and to moisten somewhat the earth with water, after the seeds be stowed in these." Meanwhile, across the English Channel, plants set out in large urns were becoming one of the trademarks of the French Renaissance garden, notably lemon and orange trees arranged in boxes. Across northern Europe, pomegranate and bay trees and vulnerable perennials were often grown in tubs, but brought under cover for the winter. And it was French gardeners who in the mid-19th century introduced the term *jardinière* for an ornamental pot or pot stand. Gertrude Jekyll, who liked to use container-grown plants to fill in seasonally empty spaces in her borders, explained helpfully: "The obvious meaning of *jardinière* is female gardener, whereas . . . it is a receptacle for holding pot plants."

And what of North America? William Faris, a craftsman clockmaker in Annapolis, Maryland, during the 18th century, recorded in his diary details of his favorite containers: earthenware pots in which he grew his special ice plants, eggplants and Jerusalem cherry trees. He dutifully noted with a watchmaker's precision: "I moved the Potts into the seller for the Winter," attending to them on a regular basis with "new dirt." He also used ordinary wooden boxes and unpainted half-barrels, in which he arrayed tuberose, wallflowers, India pinks and tulips.

Then there was Grant Thornburn, who in 1805 set up a prosperous New York seed business, and who seized upon a successful new marketing strategy: "It came into my mind to take and paint some of my common flower-pots with green varnish paint, thinking it would better suit the taste of the ladies than the common brick-bat colored ones." Two years earlier, Rosalie Stier Calvert of Riversdale, Maryland, was noting the advantages of container growing: "I have arranged all the orange trees and geraniums in pots along the north wall of the house, where they make a very pretty effect, and the geraniums, being shaded, bear many more blossoms and are growing well." This custom of creating "a pretty effect" with container plants close to the

Container plants thrive in sheltered spots such as patios and terraces and add interest to an otherwise dull area.

TOOLS IN ACTION

Using Containers to Best Effect

The style, proportion and material of a window box or container should be in keeping with the window space or other architecture of its situation. A tall container allows the plant to form a longer root system that prevents wilt in hot weather. The larger the pot, the better it is in providing adequate space for growth. Different plants can be grown next to one another in separate containers, using the appropriate growing medium for each. Acceptable results are possible in a short time, for container plants provide an instantaneous effect and it's easy to change or replace those that are weak, diseased or dead.

house continued with the Arts and Crafts-style "Craftsman" houses: an image of one such 1904 home in the New York Library collection clearly shows an entrance with a formal arrangement of large shrubs in pots.

While English gardeners had, in the 1880s, begun creating curious "rock-bed arrangements"—rockery-like cairns within which large pots could be concealed—there was nothing to rival the genuine Italian pot. When a journalist from *Country Home* magazine visited Frances Wolseley's school of lady gardeners in 1908, she reported on how "the Italian character" of the garden was emphasized with "terracotta pots, oil jars, and boxes." Some of the containers were sourced from "Mrs. Watts" village industry at Compton (*see* p. 199), "but the lemon pots and Ali Baba oil jars are obtained direct from Italy."

The tradition of container gardening continued as strong as ever throughout the 20th century. In Denmark, up to the late 1960s, Copenhagen City Council still overwintered their 10-foot high (3 m) laurel trees in large tubs in warm cellars, a practice that harked back to the Renaissance in France. It was the German Renaissance that introduced what became one of the most widespread applications of Christmas container gardening, the decorated conifer. The custom of bringing into the home a tree in a pot spread through Europe and North America from the late 1700s. None, however, could quite rival the potted myrrh trees of good Queen Hatshepsut.

Terracotta Pot

For practical and aesthetic reasons, the plain clay pot has been an enduring feature of the garden for several thousand years. It was one of the first handcrafts to go into mass production and, despite the threat of the plastic plant pot, old clays continue to grace the garden.

T erracotta has featured in gardens for almost as long as gardens have existed—4,500 years at least. Pot making is one of the oldest handcrafts and was the first to go into mass production, in Egypt around 400 B.C. Terracotta pots figured in all the great gardens: Frida Kahlo's Casa Azul in Mexico, where the artist had giant pots built into her studio walls; Thomas Jefferson's Monticello at Charlottesville, Virginia; Claude Monet's Giverny in Normandy, France; and Gertrude Jekyll's Munstead Wood in Surrey, England. Jekyll was fastidious about her supplier, preferring to patronize the pottery of Absalom Harris. (That Harris also provided the pots for Queen Victoria's royal gardens gave them a certain prestige.)

Terracotta, literally "baked earth," was used throughout the garden for border edging, French drains, roof tiles for the potting shed, rhubarb-forcing bells and, above all garden pots. It is the ideal material to contain garden plants; the semi-porous baked clay insulates plants against extremes of heat and cold. Clay pots, unlike their plastic counterparts, are also regularly recycled, the broken shards serving as drainage crocks in another pot (*see* p. 104) or, ground into granules, returned to the soil to improve drainage there.

POTS BY THE MILLION

Plastic pots, a product of the 20th-century petrochemical industry, first appeared with that great American institution, the garden center. These horticultural emporia took the gardening public by storm when they started opening in the 1950s, introducing gardeners to a host of items they never knew they needed. Some garden centers, however, have continued to feature traditional terracotta pots, boosted in part by the gardener's quest for greener, more environmentally friendly tools and equipment.

The pot makers were, once upon a time, labor-intensive local enterprises. Absalom Harris, who set up his business in Surrey, England, in 1872, dug his own clay pit, kept four kilns in constant use and employed 30 pot makers. By contrast, the Nottinghamshire pot maker Richard Sankey set up his works at Bulwell and by the early 1900s was producing half a million pots a week in kilns that swallowed up over twelve tons of coal as they baked 50,000 pots at a time. Sankey had his own railway siding and he was soon using the railroads to ship his pots abroad to British Columbia, Jamaica and New Zealand.

These pot makers used essentially the same processes,

The British pot and brick firm, Sankey of Bulwell, Nottinghamshire.

Three legs support this 2,000-year-old Mayan terracotta pot with a glazed finish.

although somewhat industrialized, as Mesoamerican potters had done 4,000 years earlier. The clay was dug out, washed and weathered for a month or two before being watered to a malleable consistency. It could then be shaped by hand, thrown on the potter's wheel, or moulded before being sun-dried and kiln-fired.

Curiously, terracotta enjoyed a renaissance on both the American and European continents around the same time. Two thousand years ago, terracotta was heavily used by both the Mayan and Roman cultures. While Mayan potters mostly served their local markets, Mediterranean potters exported their products across northern Europe as far as England's border with Scotland —the remains of these terracotta pots became a key tool in the understanding and dating of archaeological sites in later years. While the ceramic crafts continued to evolve and develop more sophisticated forms, especially in Asia, the bread-and-butter business of garden-pot making has continued to the present day.

The great advantage of terracotta was that, once a mold had been made, pots could be turned out in great numbers. One especially fashionable design in the 19th century was based on the Townley Vase, named after Charles Townley, the British collector who had bought it from its finder, Gavin Hamilton, in 1774 for £250. Hamilton had unearthed the find at an archaeological site southeast of Rome in 1773. It later passed to the British Museum.

Aside from the cultured Townley Vase, there were Oxford vases (classical-shaped with overscrolled handles), copies of older Georgian urns, and scroll pots. Scroll pots were sometimes dubbed "Jekyll pots" because she employed them so often in her designs. Jekyll did not, however, champion the benefits of terracotta: "There can scarcely be a doubt that the happiest material for our garden sculpture and ornament is lead," she declared in *Garden Ornament* (1918).

Companies like Harris and Sankey were quick to capitalize on the burgeoning 19th-

Plants in different stages of development are set out in plain and decorative terracotta pots along a terrace.

century market for terracotta. One of the foremost was that of John Marriot Blashfield. He had moved his London business to Wharf Road, Stamford, in Lincolnshire because of its better-quality regional clays, and when the first kiln was fired up in March, 1859, he invited the Marchioness of Exeter to officiate at the ceremony. (The astute Blashfield had included a model of Queen Victoria in his kiln, the figure being presented to the Queen later by Prince Albert.) Blashfield displayed his garden terracotta in a large showroom (his stock of 1,400 items, often stamped "J. M. Blashfield, Stamford Pottery, Stamford," included chimney pots, balusters, trusses and vases) and he took prizes for his work at the Great Exhibition in 1851 and the Paris world fair in 1867. Although the business failed in the 1870s, many of his ideas were exported to America through his employees (see p. 201).

THE FINE-ART POT

Not everyone admired this drive for mass production: the artist and founder of the Arts and Crafts movement, William Morris, reportedly refused to enter the hallowed halls of the Great Exhibition when he was taken there as a 17-year-old because of the industrial nature of so many exhibits. Thirty years later he would write: "Have nothing in your houses that you do not know to be useful, or believe to be beautiful." Sentiments like these, and the grinding poverty that accompanied the Industrial Revolution,

gave rise to many projects designed for the "betterment" of the working classes.

One of these, the Compton Potters' Art Guild, became so renowned that it won medals at the prestigious Chelsea Flower Show and commissions from designers and architects such as Edwin Lutyens and Clough Williams-Ellis,

TOOLS IN ACTION
Terracotta Practicalities

The person who held the lowest position in the pecking order of the Victorian garden was the pot boy, the lad responsible for cleaning and preparing the clays for the undergardeners. He was taught how to sterilize the pots in the autumn and store them over the winter in the potting shed. The old techniques are still practicable. Clay pots can be scrubbed with an organic sterilizing solution made with white vinegar or bleached in a solution of one part bleach to ten parts water. Alternatively, clay pots can be lightly baked, placed in an oven for an hour at a low 220°F (104°C), to kill any fungal growth on the clay. Salt- or hard-water deposits can be removed with a scrubbing paste of baking soda in water. Pots should be stored upside down, with some rag cushioning if they are stacked.

Next to the arbour near the plaster lady rose a sort of shed made of logs.
Pécuchet kept his tools there, and passed delightful hours shelling seeds,
writing labels, and putting his little pots in order. He would rest sitting on
a box in front of the door, contemplating improvements in his garden.

Gustave Flaubert, *Bouvard et Pécuchet* (1881)

the designer of North Wales's bizarre Italianate
village, Portmeirion. The guild was established
by a Scottish artist, Mary Seton Watts, in Surrey
as a memorial to her late husband, the painter
and sculptor George Frederic Watts. Mary had
gathered together the villagers of Compton in the
1890s to work on a mortuary chapel. She had
championed the revival of the Celtic style (a
Celtic weave was a favorite decorative style for
terracotta pots) and she was commissioned to
produce designs by the fashionable Liberty and
Co. of London. Liberty then began commissioning
Arts and Crafts-style terracotta pots (including
the popular snake-pot style) from the Potters'
Art Guild in the early years of the 20th century.
By now other ceramic companies were moving
into the mass-production of terracotta, including
the prestigious Royal Doulton Company. In the
1900s, however, the company helped fuel an
already heated horticultural dispute about
terracotta garden gnomes.

Companies such as Doulton had begun
producing popular ceramic ornaments based on
the detailed drawings of "grotesque dwarves" by
the 17th-century engraver Jacques Callot. Figures
made of terracotta, such as the famous mother-
goddess figurines discovered in the Indus Valley
of modern-day Pakistan, date back thousands of
years. But the German *Gartenzwerge* ("garden
dwarves") that flooded into America, France and
other parts of Europe were of more recent origin.
One of the earliest introductions was by an
English aristocrat, Sir Charles Isham, who placed
a group of garden gnomes he had purchased in
Germany in the gardens of his Northampton
country seat, Lamport Hall, in the 1840s. The
craze for garden gnomes sent a shudder of
apprehension through the gardening world and
The Royal Horticultural Society moved quickly
to ban "brightly colored mythical creatures" from
its annual Chelsea Flower Show. The ban was
controversially
lifted in 2013.

Stoneware Urn

The advent of stoneware gave the gardener a practical and virtually weatherproof material for utilitarian containers, decorative statuary and the classic Greek urn. The 18th-century obsession with the classical garden produced a rash of horticultural stoneware.

Definition

A vase-shaped container of classical inspiration, used as a posh planter by gardeners from the 18th century onwards.

Origin

Stoneware is a conglomerate of clay, crushed flint and glass, moulded to shape before being fired at high temperatures.

S ome of our contemporary tools originated in Greek and Roman times. The discoveries of old Roman statues immaculately preserved in volcanic ash at Herculaneum near Mount Vesuvius and, in 1748, at nearby Pompeii, whetted the aristocratic appetite for classical pedestals, neo-classical statuary and the iconic urn (from the Latin *urna*, related to *urere* "to burn," as being made of burned clay). Real hand-carved stone was expensive and lay out of the reach of most gardeners, but this did not diminish the demand for classical reproductions. The time was just right for an astute London businesswoman to meet that demand.

"Mere Makeshift"

Manufacturers on both sides of the Atlantic had started using stoneware in the late 18th and early 19th centuries to replicate features of the classical garden. There was a growing interest in stoneware birdbaths and fountains, cherubs and maidens, wellheads and planters, and Daniel Pincot set up his artificial stone factory in Lambeth, London, to cash in on the demand. The factory was acquired by the remarkable Mrs. Eleanor Coade in 1769. By 1784, when she was 51, she could afford to spend a considerable sum on a facelift of her new country home, Belmont, in Lyme Regis, Dorset.

Mrs Coade had taken over Pincot's stoneware factory opposite London's Houses of Parliament in Lambeth to make decorative and utilitarian garden articles from artificial stone. Neither her gender nor the business climate were in her favor: women were mostly excluded from business matters and the market for artificial stone was uncertain.

Yet Coade's early catalogs were soon listing fountains, figurines and gate piers alongside Medici vases, mermaids, sphinxes, nymphs and lions. She began to supply stoneware to some of the great garden designers and architects of her day—eminent men including John Nash, who had rebuilt Buckingham Palace, the designers Robert and James Adam and the architect Sir John Soane.

What were the ingredients that went into Coade stone? Eleanor, who declared it a trade secret and promised to take the formula with her to the grave, merely described it as an artificial stone known as *lithodipyra*, from the Greek *lithos* (stone), *di* (twice) and *pyra* (fire).

In reality, lithodipyra was no more than a mixture of pre-fired clay, sand, flint and cullet (crushed glass). Once combined, the mixture could be poured into any mold and then fired in a kiln at such high temperatures that it vitrified into a material with a ceramic-like texture and the strength of stone. Eleanor Coade may have concealed the true nature of lithodipyra as a tax dodge: the British treasury had imposed a levy on brickmaking in 1784, and Coade stone, as a clay-based product, might have been subject to the tax.

After she died in 1821, Eleanor Coade's business was bankrupted by the Duke of York, a notorious debtor who failed to pay his bill in time. Many of the factory moulds were sold to a former employee, Mark Blanchard, who carried on the stoneware business. He took display space for his stoneware at the Great Exhibition of 1851 in London, inspiring competitors such as the pottery owner and artist, John Marriot Blashfield. Coade, Blanchard and Blashfield stoneware would all one day fetch high prices in the garden antiques market.

After Blashfield set up business in Lincolnshire, his adopted son Joseph Joiner and foreman James Taylor left for America. Joiner worked for the New York Architectural

TOOLS IN ACTION
Stoneware Care

The popularity of stoneware on both sides of the Atlantic resulted in the manufacture of hundreds of thousands of pieces (and millions of reproductions in other materials). Real stoneware is resistant to frost, but still subject to physical damage and discoloration. Many gardeners prefer the weathered look for stoneware; however, composition stone can be cleaned with plain, warm soapy water and a soft brush. An old toothbrush is useful for cleaning inside tight nooks and crannies. Original pieces from the Coade, Blanchard and Blashfield factories bear a discreet stamp somewhere on the base, and they are valuable; practical pieces such as planters are best kept purely as decorative items.

Terracotta company and Taylor joined the Chicago Terracotta company on Long Island. "The works at Long Island City have furnished designs for more than two thousand buildings scattered through the principal cities of the Union," reported *Popular Science* magazine in 1892. Taylor was hailed as the father of American terracotta. Coade's stoneware, nevertheless, was regarded as far superior to terracotta.

A Medici stoneware urn at Kew Gardens, one of the products of Eleanor Coade's 18th century London factory.

Wheelbarrow

Ornamenting the unused wheelbarrow with flowers has become something of a gardening cliché. Thomas Jefferson advocated the use of a two-wheeled version, but the single-wheeled barrow has been a popular tool ever since its invention almost 2,000 years ago.

Definition

A single- or double-wheeled handcart for transporting material around the garden.

Origin

A Chinese invention, the wheelbarrow reached Europe some time in the Middle Ages.

When the Belgian painter Clémence van den Broeck, born in 1843, first exhibited *The Gardener's Barrow*, he pictured a giant timber affair brimming with flowers and fitted with raked back legs, handles as long as the gardeners' legs and an iron-shod wheel. But the first barrows were rather different.

THE WOODEN OX

The historical origins of the barrow hark back to a battle in western China around 1,700 years ago. This was a time when the great strategist Zhuge Liang (A.D. 181–234), a minister in the province of Shu Han, was writing his guide to warfare, *Mastering the Art of War*. The man they nick-named *Wolong* or Crouching Dragon was wrestling with the problem of provisioning an army where the local terrain was as hilly as it was muddy and labor was in short supply. His solution was the Wooden Ox, or wheelbarrow.

Zhuge's barrow was equipped with a single large wheel, over 3 ft. 3in (1 m) in diameter, which projected through the center of a carrying platform. Loads were arranged around the wheel

A wheelbarrow, first used around 1,700 years ago, could double the amount of material carried by the gardener.

and the laborer, using his or her own weight to counterbalance the cargo, pushed from behind. The wheel was narrow and could cut through the mire like a knife through butter. On the other hand, a hidden rock would tip the barrow and its contents over, a design fault that was addressed later with a barrow designed to be pulled rather than pushed.

Research seems to place Zhuge's invention around A.D. 231: was it a world first? Probably not, although the lack of evidence is more likely due to a lack of records rather than native ingenuity.

Workers were certainly wheeling their barrows around in Europe by medieval times: King Henry III's men were employing them in 1222 during building work at Dover, while a wheelbarrow was depicted in a stained glass window at Chartres Cathedral in France, dated around 1220. (A pillar carving in the cathedral is also said to picture the Ark of the Covenant being carted along on a barrow-like vehicle.)

Since the wheelbarrow was capable of doubling any porter's carrying capacity, the concept was quick to catch on. Less popular was the sail-assisted version, noted by a Dutch-American businessman when he visited China

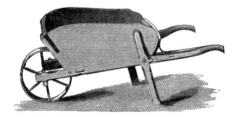

The Chinese carriers of these heavily loaded wheelbarrows could, with their sails, take advantage of the wind.

in the 1790s. This was Andreas Everardus van Braam Houckgeest who, after the failure of an earlier British trade delegation to China (*see* p. 23), enjoyed a more successful encounter with the Qianlong emperor.

Given the freedom to travel the country, Houckgeest was struck by the strange sight of wheelbarrows assisted by "a sail, made of matting or more often cloth . . . just as on a Chinese ship."

The Chinese barrow saw more horticultural action than ever before during the construction of the Yuan Ming Yuan (Gardens of Perfect Brightness), the gardens of the Old Summer Palace outside Beijing, between 1736 and 1795. It was a huge undertaking, ordered by China's Qianlong emperor and designed to recreate dozens of China's famous vistas, with lakes, pools, streams and exotic buildings and pavilions. The exclusive preserve of the imperial family, the Summer Palace was sacked, looted

and burned by British and French forces in 1860 in what the French writer Victor Hugo called a "barbarous act of robbery."

In the late 1700s, President Thomas Jefferson noted how, on his Monticello estate near Charlottesville, Virginia, a two-wheeled barrow could carry four times the load of a single-wheeled one, or "75 bricks with the old lime sticking to them." He timed his brick mason, Julius Shard, filling the barrow and carrying the load thirty yards (27 m) in less than two minutes. Maybe because Jefferson did not have to do the pushing himself, the two-wheeled barrow was quietly consigned to history.

The wooden wheelbarrow continued to evolve until the Industrial Revolution and the new metal age, which saw the introduction of wrought-iron wheels and steel or galvanized bodies: "Mounted on strong single iron frame, fitted with unbreakable steel wheel," boasted one 19th-century manufacturer's catalog.

Up until then, the barrow had been the product of several cottage craftsmen including the sled maker and the wheelwright. Sleds, sledges, sleighs or the Alpine *luge* (handy for heaving a couple of milk churns down the hillside) were common enough in snowy conditions, but small sleds were regularly used to cart lightweight loads around orchards and gardens. Construction fell to the village carpenter, who would attach a couple of rails to a ladder arrangement of cross members, with hemp cords or handles at the front and back.

It was handy to pull and push a sled at the same time. The carpenter also built wheelbarrow bodies, while the wheelwright made the barrow wheel.

That observant commentator on the domestic scene, Mrs. Beeton, regarded the barrow as the "first, and most important [tool of] all." Addressing readers of her *Garden Management* of 1861, she declared it "the conveyance of mould, manure, weeds, litter, etc., from one part of the garden to the other, and from the stable yard and manure heap to the garden." By then Mrs. Beeton's contemporaries in the English Midlands foundries were mass-producing and marketing every type of garden, fodder, offal and ash barrow ("the top can be taken off, making an Excellent Leaf and Garden Barrow"), such as the "classic No 4 Strong Garden Wheelbarrow with ash frame, deal body, wrought-iron wheel painted; (shifting boards for carrying leaves, 12/- extra)."

The major design change that lay ahead concerned that harsh-riding wrought-iron wheel. Once seeds of *Hevea brasiliensis*, the Brazilian rubber tree, had been dispatched to London's Kew Gardens in 1875, and then sent on as plants to create the great rubber plantations in Malaysia and Indonesia, the falling price of

The metal-wheeled barrow was rendered obsolete by the advent of the rubber wheel.

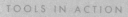

Tire Repair

One unsustainable act of gardening is to dump the garden wheelbarrow just because it has picked up a thorn and a flat tire. Most auto tire shops can repair a flat, or install a tubeless for around the quarter of the cost of a new barrow. To avoid transporting a wheelbarrow with a flat to the tire store, turn the barrow on its back and use a pair of wrenches to loosen the nuts holding the axle in place. (Use a penetrating oil if the nuts are stiff.) Alternatively a tire sealant can be employed—but the source of the puncture, such as a thorn, must first be removed. With the right tools, anyone familiar with changing and repairing a bike tire can repair a flat. Keep tires well inflated and consider replacing the pneumatic tire with a flat-free, solid-rubber one.

rubber tires threatened to put the wheelwright out of business. Later, it was an innovative vacuum-cleaner inventor, James Dyson, who came up with an alternative to Crouching Dragon's Wooden Ox in 1974: a load-bearing, pneumatic barrow with a giant ball for a wheel—the ballbarrow.

Patio Brush

New additions to the tool shed fall into the same category as satisfied gardeners: rare. One newcomer is the block paving brush, a device that arrived in response to the gardener's desire to replicate a particular feature favored by the medieval Muslims who occupied Al-Andalus: the patio.

Definition

A brush designed to keep paths clean and patios free of weeds.

Origin

The block paving brush is a recent introduction to the gardener's tool shed.

T he block paving and patio weed brush, sometimes equipped with a blade designed to uproot particularly tough weeds, was developed to help clean up patios and paved areas in the garden. It was an improvement on the retired cutlery raided by the patient soul described by the poet Rudyard Kipling as "grubbing weeds from gravel paths with broken kitchen-knives" (*The Glory of the Garden*, 1911).

Despite the relatively recent provenance of the paving brush, the patio itself was brought to prominence almost 13 centuries ago, in the *pueblos blancos*, the whitewashed villages of Al-Andalus, or Andalucia. The high plains and sierra of Spain and Portugal, occupied by

Muslims from North Africa for almost eight centuries until 1492, were graced with Islamic art and architecture. A central feature of Moorish domestic life was the courtyard or inner patio, surrounded by cool arcades and focused on the reservoir or *alberca* decorated with glazed ceramic tiles (*azulejos*). The Spanish Moors were driven out in the 15th century and they left behind outstanding examples of Islamic gardens in the royal palace of Alhambra, with their cool enclosed patio and mirror-like pools.

The paved courtyards of the Alhambra were very fine, but when the patio was translated to more northerly latitudes it suffered the vagaries of a colder climate. The inevitable build-up of weed, moss and algae led to the development of the patio brush, a tool that turned out to have useful applications elsewhere in the garden, specifically on paths and pavements.

Garden paths, together with walls and hedges, have, from André Le Nôtre's Versailles to Vita Sackville-West's Sissinghurst, traditionally been used to provide the structure of a garden. Francis Bacon would have his alleys "spacious and fair," while Thomas Hill advised that paths be well drained and of such a width that the gardener could more weed between the beds.

The patio was brought to prominence on a grand scale at the Alhambra Palace in Granada, Spain.

SOLVITUR AMBULANDO

For the poet William Wordsworth the paths at Dove Cottage, his little retreat near Grasmere in England's Lakes District, were as important as the flagstone floors inside. In 1807 he wrote to Lady Beaumont explaining how the paths or "alleys" made this garden "a place of comfort and pleasure from the fall of the leaf to its return—nearly half the year." One of the Romantic poets, Wordsworth spent hours wandering in his garden in search of inspiration—*Solvitur ambulando* (It is sorted out through walking), as the Latin proverb put it.

In more modern times, Lawrence D. Hills in his classic *Organic Gardening* (1977) advocated a central path linking the traditional four vegetable beds for the average semi-detached (i.e., duplex) house. His description evoked the suburban back garden of the 1950s with its "cropping space, bush fruit, herb bed near the kitchen door, tool shed and compost bin behind the garage, and the lawn crossed by the clothes line."

The garden path also quartered the vegetable garden (harking back to the ancient convention of dividing the garden into "quarters"), edged with low, clipped hedges that were as central to the gardens of ancient Persia as they were to gardeners settling into their new homes in the Surrey suburbs during the 1930s. The vexing issue was how to keep the paths clean. "Do not hoe up paths as this renders the garden unsightly," instructed one 1930 garden guide, advocating instead an effective weed killer such as copper

Paths knit the different elements of the garden together. Those made from smooth paving tiles are easily swept clean with a broom.

sulphate or caustic soda. The more environmentally friendly paving brush had yet to enter the scene.

The path connected the three essential elements of the garden. William Shenstone, an early English landscape gardener, described these elements as the "species" of the garden, listing them as "kitchen-gardening, parterre-gardening and landskip, or picturesque-gardening, which . . . consist of pleasing the imagination by scenes of grandeur, beauty or variety" (*Unconnected Thoughts*, 1764). The winding path, or what the garden designer Lancelot "Capability" Brown described as the "sinuous line of grace," connected all, leading to some borrowed view or feature of which the gardener was especially proud.

The path's direction of travel was a theme for horticultural discussion. The 19th-century American garden designer Andrew Jackson Downing preferred a winding path. The examples he explained in *A Treatise on the Theory and*

Maintaining a Path

Just as a garden wall needs "a strong pair of boots and a waterproof hat" (reasonable foundations and weather protection), a garden path requires a well-drained base (such as finely broken bricks—although, on well-drained ground, paths can be laid on sand) and a slight list to starboard (1 in 100) to throw off standing water. If the path has been dry-laid (that is, without mortar between the pavers or bricks), this allows the gardener to plant creeping flowers and herbs in the crevices. The inevitable weeds that also colonize the cracks can be kept in check with a paving brush.

A rough cobbled path snaking through a flower bed gives added shape and interest to a garden.

Practice of Landscape Gardening (1841) were to be found in the "cottage residence of Washington," referring to the essayist Washington Irving. George Washington also preferred "serpentine walks" at his Mount Vernon garden in Fairfax County, Virginia. Washington had been influenced by an earlier author and garden designer, Batty Langley, who in *New Principles of Gardening* (1728) had expressed his disapproval of "stiff regular Gardens," advocating serpentine walks as a more pleasing alternative. Earlier still, in the 1630s, China's Yüan Yeh detailed how paths might be made from brick waste "laid out in a path of cracked ice." Frances Wolseley liked cottage garden paths with "pansies springing up between an old paved path," a concept she thought could be adopted in larger gardens.

The materials used for paths and patios ranged from interlocking paver to river pebbles and Roman mosaics. Mediterranean gardeners substituted waste olive stones for gravel; Cornish gardeners laid local slate as paving stones, while any community with a local brickworks turned the reject bricks into paths. James Shirley Hibberd recommended in *Profitable Gardening* "no permanent walks, but narrow alleys, only trodden with the foot and every year turned over." This only produced a muddy mire in winter, and led the editors of the 1930s *Illustrated Gardening Encyclopaedia* to suggest "a combination of square paving with crazy is very effective." It was time to reach for the paving brush.

Sundial

Modern mass-produced sundials are often neither useful nor attractive. Although the older and "finer" shadow clock provides no more than an approximate time check, it stands as a constant reminder that time, even in the garden, moves on.

F unctional and decorative sundials have graced gardens for thousands of years. Sundials must have featured in the Hanging Gardens of Babylon, if those wondrous gardens did indeed exist. Billed as one of the Seven Wonders of the Ancient World, the garden, built by King Nebuchadnezzar II more than 2,000 years ago, was likely one or more ziggurats lushly planted with palms, pines and vines. Since the king's servants included some masterful mathematicians and dialists, the gardens were surely adorned with the finest examples of their craft. The Hanging Gardens certainly impressed the Greek author Diodorus Siculus, who provided detailed descriptions of them in his *Bibliotheca Historica*. But Diodorus was writing more than 500 years after Nebuchadnezzar. If there are any archaeological remains of the gardens, they have yet to be found.

The Babylonian dialists were followed by the Egyptians and the Greeks, and both civilizations made their own contributions to the science of the sun clock. The Greek Andronicus of Cyrrhus built the Tower of the Winds, which included sundials in its design, in around 50 B.C.

The octagonal Tower of the Winds in Athens served as a sophisticated timepiece over two thousand years ago.

Tyme passeth and speketh not
Deth cometh and warneth not
Amende today and slack not.
Tomorrow the self cannot.

Inscribed on a stone sundial made by
Philip Jones, Herefordshire [1600s]

In terms of history, it was the formal gardens of the 16th century that saw the shadow clock enjoy a surge of popularity, nowhere more so than in the traditional knot garden. The figure of the knot, a significant feature of Roman, Islamic, Celtic and Christian art, was incorporated into many 16th- and 17th-century garden designs. A sundial (or some similar architectural feature) was used as the centerpiece of the knot, the strands of which were formed from low-growing hedges of box, clipped rosemary or thyme. The spaces between were sometimes filled with flowers.

Sometimes the knot was planted in the shape of the lemniscate or infinity symbol ∞, introduced to the scientific world by the English mathematician John Wallis in 1655. With a sundial or an armillary sphere (its metal circles indicated the movement of the celestial spheres) as its centerpiece, the knot garden, like the Japanese Zen garden (*see* p. 76), became a place for restful contemplation. By the mid-17th century, while the knot developed into more open, looser designs, sundials remained popular.

TOOLS IN ACTION
Setting up a Sundial

Because the earth turns on a tilted axis, and because its orbit is elliptical, the shadow path of the sun varies according to the time of the year. This means that in solar time the shadow cast by the gnomon will alter at different times of the year. In order to tell the time accurately, the sundial must be set to true north (or true south in the southern hemisphere). The gnomon must also be adjusted to the latitude. The sundial reflects the saying that gardening is an inexact science: in the southern-hemisphere summer, the discrepancy between the gardener's watch and the shadow clock can be up to half an hour. In the north the error is less.

Like the parish church or stable clock, they helped the gardener to keep track of time. They still did in 1903, when Helena Rutherford Ely noted in *A Woman's Hardy Garden*: "For a time after my sun-dial was set, it was amusing to notice how often, about half an hour after eleven o clock, and again at five, this late addition to the garden would claim the attention of the workmen."

In its simplest form the sundial consists of a gnomon or triangular pointer set on a horizontal dial engraved with the hours. As the sun casts its shadow the gnomon indicates the hours and quarters. The Indian maharaja Jai Singh II had a monster version constructed in his garden of observation devices at Jaipur in the 1700s, although it was no match for Santiago Calatrava's giant Sundial Bridge, built in 2004 in Turtle Bay, California, with a 217 ft. (66 m) gnomon.

Dials might be set vertically on a sunny wall, but most are horizontal, supported by a pedestal. (Older sundials have rarely managed to remain united with their original pedestals.) Dialists, who tended to come from the clock-making and optical professions, were proud enough of their products to have their name inscribed, often on the underside of the device.

The 20th century saw a rush of sundial reproductions marked with kitsch quotes about the passing of time such as *Tempus fugit* (Time flies) or *Zähl die heitren Stunden nur* (Count only the sunny hours). Most of the reproductions are, however, hopeless timekeepers.

The sundial served to remind the gardener not only of the passing of time, but of the brevity of life.

Hose

There is a practical purpose behind the cool pools that grace famous gardens such as those of India's Taj Mahal or the Alhambra Palace in Granada: they are a source of garden water. The arrival of the rubber hose in the 1880s, however, promised to transform the gardener's world.

Definition

A flexible tube, usually fitted with a male connector at one end and a female connector at the other, used to transport water.

Origin

From the Dutch *hoos* (with cognates in Danish, Norwegian and Swedish), "a snake."

I n the past, hot climates, sparse rainfall and the need to keep food plants alive required clever and complicated systems of irrigation. From open earth ditches to carved lindenstone canals, and from open rills formed from upturned terracotta roof tiles to reservoirs and aqueducts, the peoples of the Middle East and the Mediterranean, the Persians, Egyptians, Romans and Moors all developed sophisticated and effective methods of transporting water around the garden.

However, the technology available for supplying water could be extended beyond vegetables and herbs, and wealthy landowners sometimes employed their irrigation systems in more frivolous ways. At the Villa di Castello in Tuscany in the 16th century, for example, the headstrong and despotic Cosimo I de' Medici installed a system of bronze pipes hidden under the paving, solely for the purpose of *giochi d'acqua*, water games. He could, at the turn of a key, instantly drench unwary visitors.

FIREFIGHTING

Still in the 16th century, Thomas Hill showed different methods of irrigation, including the use of a pump and water from a wooden trough, in *The Gardener's Labyrinth*. He explained "the commended times for watring of the garden beddes, and what maner of watre ought necessarily be used to plants, with the later inventions of diverse vessels aptest for this purpose." The

TOOLS IN ACTION

Storing Hoses

In the long run a cheap hose is no saving at all, since it is apt to twist, kink and develop leaks. Go for a quality reinforced vinyl or natural rubber product. Spiral hoses or those with an outer casing of thermoplastic rubber are kink-resistant and more expensive, but they are made to last. It helps if a hose, when not in use, is kept uncoiled and as straight as possible. This will minimise the occurrence of kinks. If a hose is rolled or wound on a reel, this can greatly contribute to twists and bends. Stretch it out along the edge of a lawn or along a pathway.

17th-century diarist John Evelyn also illustrated "diverse vessels" with which Thomas Hill would have been familiar. They include a water tank beside which stood a barrel on a low, four-wheeled trolley. Equipped with handles for pumping, it looks like an early form of firefighting apparatus.

Water was drawn from wells, cisterns, tanks and dipping ponds, usually positioned centrally in a garden to make access and distribution easier, and carried in watering cans (*see* p. 218). Water was also delivered to the thirsty vegetable garden and glasshouses by water carriers or barrows. Later, galvanized barrels with a pouring lip and a tilting mechanism were mounted on metal frames that could be wheeled about the garden. (A modern version of this features a thirty-gallon [110 l] water carrier mounted on a wheeled frame with pneumatic tires. Like its 19th-century predecessor, the modern version is galvanized, has a convenient handle and spout and, for easy pouring, it also pivots on its frame. Plastic versions of the same capacity are available with hose attachments.)

But it was the world of firefighting, in particular that of a city hall fire, that changed the business of irrigating the garden. When Amsterdam's city hall burnt down in 1652, the accident left a lasting impression on one of the witnesses, 12-year-old Jan van der Heijden. Some 20 years later, having become an accomplished draftsman and painter, Van der Heijden was also gripped by the technical potential of various inventions. He is credited with contriving the arrangement of a water-carrying linen tube attached to a fire appliance. Later, leather superseded linen as the preferred fabric of the hose, and lengths of leather pipes were hand-stitched in 50 ft. (15 m) lengths by sailmakers. Canvas hoses sealed with tar also made an appearance, but it was not long before Van der Heijden's invention was seized upon by gardeners who recognized its potential.

In 1845 the strangely named Gutta-Percha Company started producing truly flexible hoses in the UK. Gutta-percha was a type of natural rubber, formed from the gum of a tree native to Malaysia. More significantly, it behaved like a

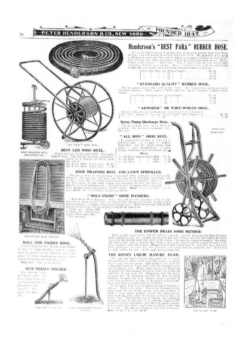

Cheap and affordable rubber hoses meant more parts of the garden could be put under cultivation.

thermoplastic material, solidifying when cool. The Gutta-Percha Company started manufacturing one-hundred-yard (91 m) lengths of hose; the tubing is said to have been modeled on Italian pasta-making machines. As so often happens, supplies of natural gutta-percha gradually dried up and collapsed because of over-harvesting and it was left to Charles Goodyear and his vulcanized rubber (*see* p. 39) to fill the gap.

From the 1870s onwards, B. F. Goodrich in Akron, Ohio, had been manufacturing flexible, watertight rubber hoses with cotton-ply reinforcement for firefighting. These soon gained a foothold in American gardens. The rubber hose did not, however, meet with everyone's approval. Edward Luckhurst, writing to the English *Journal of Horticulture, Cottage Gardener and Home Farmer* in 1883, sparked a debate about the pros and cons of the fabrication of the garden hose. "Enough 2-ply indiarubber hose in lengths of 60 feet [18 m] was procured to reach every part of the garden, with suitable brass unions for screwing together and upon the hydrants, and a copper pipe with a tap, jet, and rose for the watering," he wrote. After a couple of seasons' wear, however, "the hose was cracked and split in several places, and the conviction grew upon me that the extra 10d per foot which would have purchased the leather hose would have been a wise outlay at first."

A correspondent who signed himself T. W. S. of Lee responded, prompted by a spell of dry weather and the need to replace a worn-out hose. "We at once decided on obtaining a leather hose similar to that recommended by Mr. Luckhurst. Of course, a leather hose is more costly than an indiarubber one, but . . . the additional outlay incurred will ensure a more efficient and durable article, and afford greater satisfaction to employer and gardener."

The rubber hose would eventually triumph over its leather counterpart and has gone on to have some unexpected applications in the garden, from ventilating hedgehog hibernation houses, protecting wire ties from chafing planted trees, sheathing the teeth of the pruning saw, and even, when well-warmed and supple, acting as a stencil for marking the outline of a new flowerbed or pond.

The garden hose was central to the world's first comedy film, titled *L'Arroseur arrosé* (*The Waterer Watered*) dating from 1895 by Louis Lumière. The 49-second work features Louis' own gardener and an apprentice carpenter who plays a practical joke on the gardener, standing on the hose as the man waters his plants, thereby stopping the flow of water.

Technological advances, especially in the world of plastics, have left gardeners faced with a bewildering variety of hoses offering a battery of features, ranging from reinforced rubber or vinyl five-ply hoses to versions that are kink-resistant, UV-resistant, abrasion-resistant, fungus-resistant, cadmium-free, barium-free, leakproof, crushproof, or designed as "weepers" for drip-irrigation systems.

Slapstick comedy with a hose was immortalized by Louis and Auguste Lumière's L'Arroseur arrosé *(The Waterer Watered).*

Watering Can

If one tool can be said to epitomize the gardener's world, it is the watering can. From the elegant copper "conservatory" model to a simple galvanized one, the classic watering can symbolizes the very art of gardening.

Definition

A container with spout and handle, developed as a means of carrying water to the different parts of the garden.

Origin

The watering can evolved from the "common water-potte" during the 17th century.

The well-balanced watering can is a joy to use. Most gardeners, passing between their plants at dusk or dawn, gently tipping the swan neck to "imitate the rain falling from the Heavens," as one 18th-century commentator put it, would agree that the act of watering exemplifies all that is best about gardening. Yet what constitutes a good watering can be the cause of some dispute.

Despite being close neighbors, the French and the English cannot agree on the perfect watering can. The French gardener prefers an elegant oval-shaped, swan-necked, single-handled can that measures in liters; the English gardener is more likely to buy a stout, two-gallon, double-handled galvanized can.

John Claudius Loudon described the differences: "The French watering-pots are generally formed of copper, and some have zig-zag spouts, to break the force of the water when pouring it on plants without the use of a rose (the perforated plate at the end of the spout)." This sipping action was in contrast to the "tinned, iron or copper common watering-pot," which came in a confusing number of versions including "the common large pot, with two roses of different sizes" and the long-spouted pot for "watering plants in pots, at a small distance, either with or without a rose." Then there was the shelf watering-pot, "a small cartouche-shaped pot for watering plants on shelves." By now an ingenious Mr. Money, who probably hoped to live up to his name, had invented a watering-pot with three interchangeable spouts; Loudon judged it efficient for its "definitiveness of action."

The watering-pot was not the only device available to the 19th-century gardener. While the French favored their copper cans, the Dutch had been exporting brass cans since the 18th century. There were long, brass watering tubes, specially adapted to reach into "pine beds" (pineapple hotbeds) and potted plants placed in forcing beds, together with an array of pumps and syringes that mostly worked on the same principle as the water pistol (and were probably employed as such around the back of the potting shed). Macdougal's Inverted Garden Syringe was devised to spray the undersides of plant leaves; Siebe's Universal Garden Syringe, equipped with four interchangeable spray heads, could tackle several areas at once; while the Warner Syringe was "an

A gardener uses a long-reach brass stirrup pump to reach the back of the glasshouse staging.

The stout English-style galvanized watering can with a carrying handle on top.

imperfect substitute for the others," according to Loudon, except that "it has the advantage of being sold at little more than half their price."

There were clever contrivances called "aquarians" (a French version was described as a *soude*), water-filled, vacuum-operated buckets with a trap door or rose set in the base. Once filled, the vacuum inside the vessel held the water until it was broken by the gardener tipping a miniature lid on the top. Ideal for watering tender seedlings, it avoided drowning the plants in a downpour. The design of the aquarian originated from the early watering vessel, an earthenware container with a thumbhole at the top and a perforated base. Allowing air through the thumb hole controlled the flow of water from the base, releasing a gentle shower on the plants below. Or, as Thomas Hill explained: "The common water potte for the garden beds with us, hath a narrow neck, big belly, somewhat large bottom, and full of little holes." The vacuum arrangement "allowed the full pot to be carried in handsome manner" around the garden. Anthony Huxley in *An Illustrated History of Gardening* reports that sprinkling cans had first appeared in the 1400s and the metal versions in the 1700s: John Evelyn had depicted a curious example equipped with a pair of spouts.

By the 1880s the definitive piece of equipment, according to most observers, was an invention by one John Haws, an apparently unsuccessful planter who had been working for the British colonial service in Mauritius, endeavoring to raise what the Spanish in Mexico called the "little pod"—*vanilla*.

TOOLS IN ACTION
Watering Plants

Knowing when, and in what quantities, to water depends on knowing your soil: a sandy soil will be thirstier than a clay one. The ideal watering can is balanced so that the water's flow can be controlled by the tipping of one hand. In this way a gardener with a watering can in each hand can deliver four gallons (14 l) on each trip. As a rough guide, two trips every seven to ten days should maintain plant growth over 10 sq ft. (1 m²). New plants and prolonged dry weather will double or triple the demand for water, while wilting leaves indicate drought conditions. The ground will then require a thorough watering, ideally during the early morning or late evening, to reduce ground evaporation. The need to water is significantly reduced by mulching with 2 in. (50 mm) of organic matter, proprietary mulching sheets or gravel.

Vanilla's reliance on being pollinated by an indigenous species of bee made it notoriously difficult to grow elsewhere. Haws's efforts to raise vanilla on the little island in the Indian Ocean were further hampered by the primitive watering-pots available locally. Haw set about designing a watering can "that is much easier to carry and tip, and . . . much cleaner . . . than any other put before the public." That, at any rate was his claim.

The Haws model became a market leader. It was assisted by the efforts of John Haws's nephew Arthur who, when he took over the business, applied a sometimes obsessive attention to detail to the firm's research and development. Arthur was especially interested in the roses. Two were supplied with each can, one round, one oval, and Arthur insisted that the perforations should be spaced just so, varied in size according to the volume of the particular can, and carefully tapered in order to cope efficiently with debris in the tank. He employed a worker whose task was to punch each rose out on a treadle.

Red and green tin-plated watering cans had also begun to appear in the early 19th century. Tinplate—metal sheet coated with a rustproof layer of tin—was the plastic of the 19th century. The industry had spread from Saxony in the 1600s to Wolverley in Worcestershire and Pontypool in Wales, and by the 19th century had burgeoned into a worldwide trade, dominated by South Wales which flooded the market with

The gardening goes on: a metal watering can delivers a sustaining shower to a tray of plants.

inexpensive tin planters, garden labels and watering cans. The tinplate cans remained popular until galvanized cans, and finally plastic, replaced them.

Having purchased the perfect watering pot, the gardener had to be properly instructed on its use: John Worlidge recommended in the 17th century "the smallest or rain like drops"; Thomas Hill advocated watering into a hole, made with a dibber, beside the plant, "for water rotteth and killeth above ground"; while one Mr. Marshall advised in the 1800s: "Watering is a thing of some importance in cultivation, though not so much as many make of it."

221

A

Acland, George 71
Appert, Nicolas 163
apples 156–7
Artedi, Peter 96–7

B

Bacon, Francis 208
Baekeland, Leo 184
Balfour, Lady Eve 87–8
barrels 155–8
Bartram, John 52
basketmaking 28–31, 105
Beeton, Mrs. 206
Benyon, Benjamin 70
Bertrand de Molleville,
    Antoine 23
billhooks 72–5
bio-dynamic
    gardening 87
birds 169, 170, 171
Birdseye, Bob 162–3
Blanchard, Mark 202
Blashfield, John Marriot
    198, 202
blow torches 128
bonsai 192
boot scrapers 40
Bowles, E. S. 140
Brown, Lancelot
    "Capability" 53, 209
Budding, Edwin 115, 116
bulb planters 15–18
Burritt, M. C. 154

C

Carson, Rachel 133
catalogs 46–9
Celsius, Anders 166
Charles, Prince 88, 133
China roses 24, 25–6

cloches 182–5
cloud pruning 124
Coade, Eleanor 201–2
Cobbet, William 135–6, 162
Columella, Lucius 33–4, 90,
    169, 170
community gardens 141
composters 84–8, 106
conservatories 180
containers 190–202
Crossman, Charles 100

D

Dahlman, Karl 113
daisy grubbers 125–8
dethatching 78
dibbers 35–8
Dig for Victory 59, 185
Downing, A. J. 156
Downing, Charles
    145, 156
Durand, Peter 163
Dyson, James 206

E

Edmonds, Charles 63
Ely, Helena Rutherfurd
    175–6, 213
Evelyn, John 12, 26, 35, 51,
    52, 109, 162, 215, 220

F

Fahrenheit,
    Daniel Gabriel 166
fertilizers 134–7
Fiacre, St. 57
Fitzherbert, Anthony 61
Flaubert, Gustave 199
forks 7, 10–14
Francé-Harrar, Annie 87
fruit barrels 155–8

fruit ladders 144–6
Fukuoka, Masanobu 14

G

Galvani, Luigi 158
garden baskets 28–31
garden catalogs 46–9
garden journals 50–3
Gerard, John 69, 70
Gilbert, Dr. Joseph 136
glasshouses 180, 181
gloves 43, 44–5
gnomes 199
Goodrich, B. F. 216
Goodyear, Charles
    42, 216
grafting knives 7, 25,
    147–50
Grant, Anne MacVicar 21
glasshouses 6, 178–81
gutta-percha 216

H

hand shears 114
Harris, Absalom 196
hats 43, 45
Hatshepsut 191
Haws, Arthur 221
Haws, John 220–1
hedge-laying 73–5
hedge shears 123
Heijden, Jan van der
    215–16
herb gardens 69, 122
herbicides 37, 129, 130,
    131–3
Hibberd, James Shirley
    13, 34, 86, 90, 92, 150,
    181, 210
Hill, Thomas 34, 193, 208,
    215, 220, 221

Hills, Lawrence D.
    87, 88, 170, 209
hoes 60–4
hoses 6, 214–17
hotbeds 89–93
Howard, Albert 87–8
Howard, Arthur 82–3
Hudson, William 113–14
Huxley, Anthony 220

I

ice houses 162
Irving, Washington
    6–7, 210

J

Jackson, Andrew 145, 209
Jefferson, Thomas 50, 51,
    120, 161–2, 205
Jekyll, Gertrude 40–1, 42,
    56, 75, 128, 162, 193,
    196, 197
John Moseley and Sons 20
journals 50–3
jute 71

K

Kelley, Ruth Edna 149
Kemp, Edward 103

L

labels 159–63
ladders 144–6
Latin nomenclature 94–8
Lawes, John Bennet
    135, 136–7
lawn mowers 6, 88, 112–16
lawns 112, 114, 115,
    116, 127
Lawson, William 149
L'Écluse, Charles de 16

Liebig, Justus von 93, 136

Linnaeus, Carl 94,
96–8, 166

Loudon, Jane 19, 21–2, 63,
79, 105, 160, 167

Loudon, John Claudius 14,
16, 26, 36, 47, 66, 126–7,
160, 161, 171, 188, 219

Luckhurst, Edward
216, 217

Lutyens, Edwin 40–1

M

manure 89–93

Marconi, Guglielmo 108

Marvel, Andrew 157

mattocks 12, 65–7

measuring/measurements
139

mechanical tillers 80–3

Mollet, Claude 122

Molleville, Antoine,
Bertrand de 23, 27

Mollison, Bill 13

Morris, William 198

mowers 6, 88, 112–16

music 108, 109

N

Nicholson, Sir William 40

no-dig systems 13–14, 38

Noguchi, Isamu 77

O

Opinel, Joseph 7, 148

organic gardening 87, 88,
131, 133

organization of tools
109, 176

P

parterres 122

paths 208–10

patio brushes 207–10

Pepys, Samuel 51

permaculture 13

pesticides 36, 37, 131–3

pH scale 33, 34

Picasso, Pablo 149

plant containers
190–202

plant hunters
187–8, 189

Pliny the Elder 158,
176, 191

Pliny the Younger 114

poles 139

potting sheds
174–7, 199

pottles 30

preserving food 162–3

pruning saws 151–4

pruning shears 6, 23–7

Q

Quastel, Harry 130, 131

R

radios 107–9

raised beds 99–103

rakes 76–9

Repton, Humphry 52, 53

Retallack, Dorothy 108

Richardson, Mervyn 113

riddles 104–6

Robinson, William 16–18

rods 139, 140

roses 24, 25–6, 69

Royal Horticultural Society
24, 25, 45, 59, 189, 199

Ruskin, John 38

S

Sankey, Richard 196

scarecrows 168–71

Schwamstapper, Traugott
44, 45

scythes 118, 119–20

seed composts 105

seeds 36–8, 47–9, 59, 184

shears 121–4

sickles 12, 114, 117–20

sieves 104–6

Smith, Thomas 31

Soil Association 87

soil fertility 33–4, 87, 90–1,
92, 93, 131

soil sieves 104–6

soil-test kits 32–4

Soper, John 87

spades 7, 56–9

staffs 139

Steiner, Rudolf 87

stoneware urns 200–2

string lines 68–71, 139

sundials 211–13

T

tape measures 138–41

terracotta pots 106, 195–9

terrariums 186–9

thermometers 164–7

Thoreau, Henry David 53,
56, 57, 61–2, 140

tillers 80–3

tires 206

Timmins, Richard 19

topiaries 122–3, 124

trowels 19–22

trugs 31

Turnock, Mike 105

Tusser, Thomas 7, 34, 66

U

urns 200–2

V

Victoria, Queen 31,
180, 198

W

Walling, Edna 17, 176

Ward, Dr. Nathaniel
186, 188–9

Washington,
George 150, 162, 210

watering cans 218–21

Webb, Jane
see Loudon, Jane

weed grubbers
125–8

weed killers 129–33

Wellington boots
6, 39–42

wheelbarrows 203–6

White, Gilbert 51, 91,
169–70

wild gardens
16–18, 115

wildlife 169, 170–1

Wolseley, Frances 22,
194, 210

women 19, 20–2, 86, 114,
168, 191, 194

Wordsworth, William
209

Worlidge, John 221

Wyse-Gardner, Charles
183, 185

Z

Zhuge Liang 204

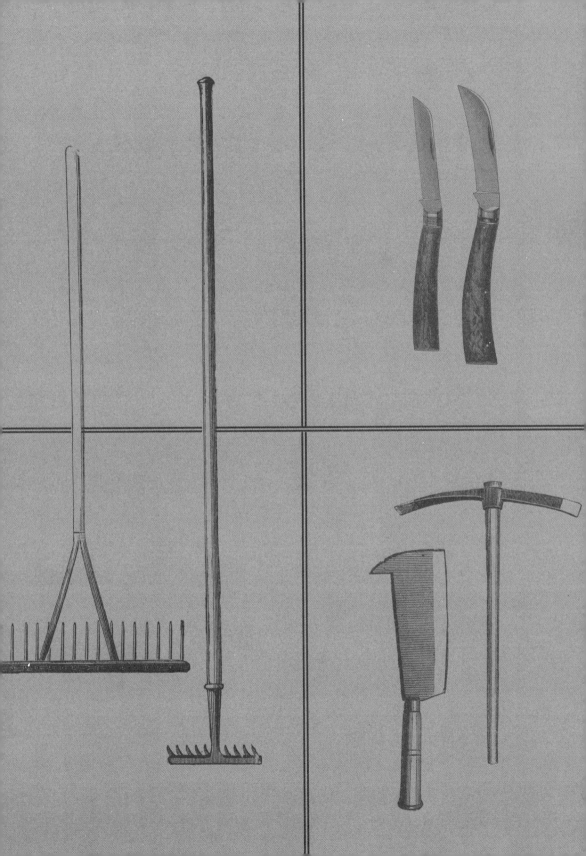